REGRESSION METHODS

STATISTICS: Textbooks and Monographs

A SERIES EDITED BY

Volume 1: The Generalized Jackknife Statistic, *H. L. Gray and W. R. Schucany*

Volume 2: Multivariate Analysis, *Anant M. Kshirsagar*

Volume 3: Statistics and Society, *Walter T. Federer*

Volume 4: Multivariate Analysis: A Selected and Abstracted Bibliography, 1957-1972, *Kocherlakota Subrahmaniam and Kathleen Subrahmaniam* (out of print)

Volume 5: Design of Experiments: A Realistic Approach, *Virgil L. Anderson and Robert A. McLean*

Volume 6: Statistical and Mathematical Aspects of Pollution Problems, *John W. Pratt*

Volume 7: Introduction to Probability and Statistics (in two parts) Part I: Probability; Part II: Statistics, *Narayan C. Giri*

Volume 8: Statistical Theory of the Analysis of Experimental Designs, *J. Ogawa*

Volume 9: Statistical Techniques in Simulation (in two parts), *Jack P. C. Kleijnen*

Volume 10: Data Quality Control and Editing, *Joseph I. Naus*

Volume 11: Cost of Living Index Numbers: Practice, Precision, and Theory, *Kali S. Banerjee*

Volume 12: Weighing Designs: For Chemistry, Medicine, Economics, Operations Research, Statistics, *Kali S. Banerjee*

Volume 13: The Search for Oil: Some Statistical Methods and Techniques, *edited by D. B. Owen*

Volume 14: Sample Size Choice: Charts for Experiments with Linear Models, *Robert E. Odeh and Martin Fox*

Volume 15: Statistical Methods for Engineers and Scientists, *Robert M. Bethea, Benjamin S. Duran, and Thomas L. Boullion*

Volume 16: Statistical Quality Control Methods, *Irving W. Burr*

Volume 17: On the History of Statistics and Probability, *edited by D. B. Owen*

Volume 18: Econometrics, *Peter Schmidt*

Volume 19: Sufficient Statistics: Selected Contributions, *Vasant S. Huzurbazar (edited by Anant M. Kshirsagar)*

Volume 20: Handbook of Statistical Distributions, *Jagdish K. Patel, C. H. Kapadia, and D. B. Owen*

Volume 21: Case Studies in Sample Design, *A. C. Rosander*

Volume 22: Pocket Book of Statistical Tables, *compiled by R. E. Odeh, D. B. Owen, Z. W. Birnbaum, and L. Fisher*

Volume 23: The Information in Contingency Tables, *D. V. Gokhale and Solomon Kullback*

Volume 24: Statistical Analysis of Reliability and Life-Testing Models: Theory and Methods, *Lee J. Bain*

Volume 25: Elementary Statistical Quality Control, *Irving W. Burr*

Volume 26: An Introduction to Probability and Statistics Using BASIC, *Richard A. Groeneveld*

Volume 27: Basic Applied Statistics, *B. L. Raktoe and J. J. Hubert*

Volume 28: A Primer in Probability, *Kathleen Subrahmaniam*

Volume 29: Random Processes: A First Look, *R. Syski*

Volume 30: Regression Methods: A Tool for Data Analysis, *Rudolf J. Freund and Paul D. Minton*

OTHER VOLUMES IN PREPARATION

REGRESSION METHODS

A Tool for Data Analysis

RUDOLF J. FREUND
Institute of Statistics
Texas A & M University
College Station, Texas

PAUL D. MINTON
Director, Institute of Statistics
Virginia Commonwealth University
Richmond, Virginia

MARCEL DEKKER, INC. New York and Basel

Library of Congress Cataloging in Publication Data

Freund, Rudolf Jakob, [Date]
Regression methods.

(Statistics, textbooks and monographs ; v. 30)
Bibliography: p.
Includes index.
1. Regression analysis. I. Minton, Paul D., [Date] joint author. II. Title.
QA278.2.F7 519.5'36 79-20458
ISBN 0-8247-6647-4

MARCEL DEKKER, INC.
270 Madison Avenue, New York, New York 10016

Current printing (last digit):
10 9 8 7 6 5 4 3 2 1

PRINTED IN THE UNITED STATES OF AMERICA

PREFACE

The purpose of this book is to instruct the reader in the usage of linear model methodology in the analysis of data. The book is organized as a textbook in that it starts with simple models and continues by discussing more-complex models and special considerations. However, the book should also serve as a useful reference for practitioners and users of statistical methodology.

Since this is a book on methodology, the coverage of theory has been kept to a minimum. A few derivations are given to provide a rationale for some concepts and associated methods; however, the examples in the text and problems at the end of chapters are intended to serve as the primary vehicle for understanding proper usage of the methodology.

The book is suitable for students who have had a basic course in statistical methodology, including concepts of statistical inference, t tests, and the analysis of variance. A knowledge of calculus or statistical theory is not required. Occasional references requiring such knowledge are included for the benefit of more-advanced users, but these references are not essential for understanding the material.

Because this is a book in analysis and methodology rather than theory, numerical examples are, of course, quite important. The advent of modern computers has reduced the heavy computing load inherent in regression problems, but this aspect still presents a problem in teaching a course. The general philosophy of the book is to present in the text relatively simple examples from small samples so as not to confuse the student with details that distract from the principles being presented. A more-thorough understanding of the applications may be obtained by means of the problems at the end of the chapters, some of which are "contrived" for easy calculation, while others are based on "live" data, where interpretation and analysis are of prime importance and provide real-life lessons in the arts that follow the techniques of statistical analysis.

Rudolf J. Freund

Paul D. Minton

CONTENTS

LIST OF FIGURES

LIST OF TABLES

REGRESSION METHODS

Chapter 1

MATRICES

1.1 INTRODUCTION

The standard language for many statistical analysis procedures is *matrix algebra*. This language is essentially a shorthand which provides for the simplification of algebraic statements involving many variables. The econometrician Valavanis (1959) expressed it in this way:

> ...and matrices, which are nothing but many-dimensional ordinary numbers -- indeed, statements that are true of ordinary numbers seldom fail for matrices -- for instance, you can add, subtract, multiply, and divide matrices analogously to numbers and, in general, handle them as if they were ordinary numbers though perhaps more fragile; a *vector* is a special kind of matrix.

The subject of matrices, including a rather extensive body of theory and applications, is the subject of many textbooks, e.g., Hohn (1964) or Horst (1963). We discuss in this chapter only certain aspects of matrix algebra that are useful for multiple regression. No proofs are presented.

1.2 DEFINITIONS AND PROPERTIES

A *matrix* is a rectangular array of elements arranged by rows and columns. This array is usually enclosed by a pair of square brackets. Each element is identified by its location in the matrix, that is, in which *row* and *column* (in that order) it is located. A matrix is usually denoted by a capital letter and the elements of the matrix by the corresponding lowercase letter with subscripts indicating location. Thus

$$A = \begin{bmatrix} a_{11} & a_{12} & a_{13} & a_{14} \\ a_{21} & a_{22} & a_{23} & a_{24} \\ a_{31} & a_{32} & a_{33} & a_{34} \end{bmatrix}$$

and in the matrix

$$W = \begin{bmatrix} 1 & -5 \\ 2 & 14 \\ 5.7 & 0 \end{bmatrix}$$

the values of the elements

$$w_{11} = 1 \qquad w_{12} = -5 \qquad \cdots$$

In general, the matrix A, say, has a typical element a_{ij}, that is, the element in the i-th row and j-th column.

A matrix is also known by its size or *order*, i.e., by the number of rows and columns it has. Thus, the matrix A (above) is a 3×4 matrix, since it has three rows and four columns. Likewise, W is a 3×2 matrix. Matrices with equal numbers of rows and columns are *square* matrices and will be seen to have special importance. A matrix having only one column is known as a *column vector* and is denoted by an underlined lowercase letter; elements in a vector need to be subscripted with only the row index. A *row vector* is a matrix with one row (a notation for this is discussed later), and a matrix with one row and column is called a *scalar* and behaves as an ordinary number or algebraic symbol.

It should be pointed out at this time that a matrix has no value in or by itself; it only provides a means for easily denoting a rectangular array of elements. There is a value which can be attached to a square matrix: this is called the *determinant*. It has virtually no use in regression.

Those elements whose row and column indices are equal (a_{ii}, say) are known as *diagonal elements*, as they lie on the *main diagonal* of the matrix.

A square matrix A such that $a_{ij} = 0$, $i > j$, that is, *nonzero* elements only above and including the main diagonal, is an (upper) *triangular matrix*. A square matrix A such that all $a_{ii} \neq 0$ and $a_{ij} = 0$, $i \neq j$, that is, nonzero elements only on the diagonal, is known as a *diagonal matrix*. A diagonal matrix A such that $a_{ii} = 1$ is known as an *identity matrix* and is denoted by I. The identity matrix has a function similar to that of the scalar one. A *null matrix* (not necessarily square) has all elements equal to zero.

A square matrix A such that $a_{ij} = a_{ji}$ is known as a *symmetric matrix*. Thus,

$$S = \begin{bmatrix} 4 & 2 & 5 \\ 2 & 7 & 9 \\ 5 & 9 & 20 \end{bmatrix}$$

is a symmetric matrix; it is easily seen to be symmetric about the main diagonal.

1.3 MATRIX OPERATIONS

Two matrices A and B are said to be *equal* only if *all* corresponding elements of A are equal to those of B. Thus,

$$A = B \quad \text{implies} \quad a_{ij} = b_{ij} \qquad \text{for all } i \text{ and } j \tag{1.1}$$

Naturally, two equal matrices must be of the same order.

The *transpose* of a matrix A of order $(r \times c)$, denoted A', is a matrix of order $(c \times r)$ such that

$$a'_{ij} = a_{ji} \tag{1.2}$$

For example, if

$$A = \begin{bmatrix} 1 & -5 \\ 2 & 2 \\ 4 & 1 \end{bmatrix}$$

then

$$A' = \begin{bmatrix} 1 & 2 & 4 \\ -5 & 2 & 1 \end{bmatrix}$$

It can be seen that the columns of A have become the rows of A'. The transpose of a column vector is a row vector; thus, we do not need a special notation for row vectors. Also, the transpose of an upper triangular matrix is a lower triangular matrix. Note further that

$$(A')' = A \tag{1.3}$$

and if A is symmetric, then

$$A' = A$$

The process of *matrix addition* is defined as follows:

$$A + B = C$$

if

$$a_{ij} + b_{ij} = c_{ij} \tag{1.4}$$

that is, an addition of corresponding elements. As an example, let

$$A = \begin{bmatrix} 1 & 2 \\ 4 & 9 \\ -5 & 4 \end{bmatrix} \qquad B = \begin{bmatrix} 4 & -2 \\ 1 & 2 \\ 5 & -6 \end{bmatrix}$$

then

$$C = A + B = \begin{bmatrix} 5 & 0 \\ 5 & 11 \\ 0 & -2 \end{bmatrix}$$

Obviously, two matrices not of the same order may not be added. The properties of matrix addition are as follows:

1. $A + B = B + A$
2. $A + (B + C) = (A + B) + C$
3. $(A + B)' = A' + B'$

Subtraction of matrices follows equivalent rules.

The process of matrix multiplication is defined as follows:

$$AB = C$$

if

$$c_{ij} = \sum_{k=1}^{n} a_{ik}b_{kj} \tag{1.5}$$

where n is the number of *columns* of A and the number of *rows* of B. For some readers it may be helpful to rephrase the process of matrix multiplication as follows:

> The element of the i-th row and j-th column of the product matrix C, c_{ij} is the pairwise *sum of products* of the corresponding elements of the i-th *row* of A and the j-th *column* of B.

In order for A and B to be conformable for multiplication, the number of columns of A must be equal to the number of rows of B. The order of the product matrix C will be equal to the number of rows of A by the number of columns of B. As an example, let

$$A = \begin{bmatrix} 2 & 1 & 6 \\ 4 & 2 & 1 \end{bmatrix} \qquad B = \begin{bmatrix} 4 & 1 & -2 \\ 1 & 5 & 4 \\ 1 & 2 & 6 \end{bmatrix}$$

Note that A has three columns and B has three rows; hence, A and B may be multiplied. The elements of C = AB are obtained as follows:

$$c_{11} = a_{11}b_{11} + a_{12}b_{21} + a_{13}b_{31}$$
$$= 2(4) + 1(1) + 6(1) = 15$$
$$c_{12} = a_{11}b_{12} + a_{12}b_{22} + a_{13}b_{32}$$
$$= 2(1) + 1(5) + 6(2) = 19$$

....

$$c_{23} = a_{21}b_{13} + a_{22}b_{23} + a_{23}b_{33}$$
$$= 4(-2) + 2(4) + 1(6) = 6$$

The entire matrix C, of order 2 × 3, is

$$C = \begin{bmatrix} 15 & 19 & 36 \\ 19 & 16 & 6 \end{bmatrix}$$

The properties of matrix multiplication are as follows:

1. Generally, $AB \neq BA$. Note that unless A and B are square, AB and BA may not even be conformable for multiplication
2. $(AB)C = A(BC)$
3. $A(B + C) = AB + AC$
4. $(AB)' = B'A'$

There is a convenient device which may help in checking whether matrices may be multiplied and also in determining the order of the final product matrix. Below each matrix indicate the order (row by column):

$$\underset{(m \times n)}{(A)} \quad \underset{(n \times p)}{(B)} \quad \underset{(p \times q)}{(C)} = \underset{(m \times q)}{D} \tag{1.6}$$

In order to be able to perform this multiplication, adjacent order indicators of the individual matrices must be the same (n and p), and the order of the product matrix will be given by the extreme indicators $(m \times q)$. Note that the product of a matrix and a column vector is a column vector, the product of a row vector and a matrix is a row vector, and the product of a row vector and a column vector is a scalar.

A special operation, called *scalar multiplication*, is defined as the multiplication of *each element* of a matrix by a constant (scalar). Note that this is *not* a special case of matrix multiplication. The results of a scalar multiplication can, however, be accomplished by a (pre- or post-) matrix multiplication using a diagonal matrix with elements equal to that of the scalar.

There is no matrix division as such; equivalent results are obtained by the use of the *inverse of a matrix*. The inverse of a *square matrix* A, denoted A^{-1}, is defined by the property:

$$AA^{-1} = I \tag{1.7}$$

Other properties of matrix inverses are as follows:

1. $AA^{-1} = A^{-1}A = I$
2. If $C = AB$ (all square), then $C^{-1} = B^{-1}A^{-1}$
3. If $B = A^{-1}$, then $B' = (A')^{-1}$
4. If A is symmetric, A^{-1} is symmetric
5. If an inverse exists, it is unique

Certain matrices do not have an inverse; such matrices are called *singular* (see Section 1.4). In a sense they play the role in matrix algebra of the scalar zero, even though they need not contain zero-valued elements. General procedures for determining the inverse of a matrix are usually quite long and tedious; they are discussed in Section 1.5.

1.4 LINEAR EQUATIONS

Manipulations involving systems of simultaneous linear equations are useful in discussions of multiple regression and are readily illustrated with the aid of the above-mentioned matrix definitions and operations.

Define A as an $m \times m$ matrix and $\underline{x}$ and $\underline{b}$ as m-element column vectors. The equation

$$A\underline{x} = \underline{b} \tag{1.8}$$

is seen to represent the system of m linear equations in m variables x_i, of the form

$$
\begin{aligned}
a_{11}x_1 + a_{12}x_2 + \cdots + a_{1m}x_m &= b_1 \\
a_{21}x_1 + a_{22}x_2 + \cdots + a_{2m}x_m &= b_2 \\
\cdots\cdots\cdots\cdots\cdots\cdots\cdots\cdots \\
a_{m1}x_1 + a_{m2}x_2 + \cdots + a_{mm}x_m &= b_m
\end{aligned}
$$

The matrix A represents the coefficients of the variables in the equations and is called the *coefficient matrix*. Such a system may be solved (for specific values of x_i) by various methods, for example by the elimination of variables using series of multiplications and additions of equations. Using matrix notation, we can symbolically solve the equivalent matrix equation by pre-multiplying both sides of the equation by A^{-1}, resulting in

$$
\begin{aligned}
A^{-1}A\underline{x} &= A^{-1}\underline{b} \\
I\underline{x} &= A^{-1}\underline{b}
\end{aligned}
$$

then

$$\underline{x} = A^{-1}\underline{b} \tag{1.9}$$

It will be seen that this exercise does not actually provide the numerical solution for $\underline{x}$, since finding A^{-1} involves as much work as solving the system of equations. However, being able to express the solution symbolically is quite helpful. Note, further, that if there are several sets of equations, say

$$
\begin{aligned}
A\underline{x} &= \underline{b} \\
A\underline{y} &= \underline{c} \\
A\underline{w} &= \underline{d}
\end{aligned}
$$

the inverse of A need be obtained only *once*, that is

$$
\begin{aligned}
\underline{x} &= A^{-1}\underline{b} \\
\underline{y} &= A^{-1}\underline{c} \\
\underline{w} &= A^{-1}\underline{d}
\end{aligned}
$$

It is well known (and intuitively reasonable) that if a solution exists, it is unique; matrix theory confirms this by the fact of the uniqueness of the inverse. Intuition also suggests that more equations than variables or *vice versa* will produce problems; this is supported by the fact that unique inverses do not exist for the corresponding rectangular coefficient matrices.

Actually, the concept of singular matrices is related to the correspondence between the number of variables and the number of equations. As an example, consider the following system of equations:

$$5x_1 + 2x_2 + x_3 = 10$$

$$x_1 + x_2 - 2x_3 = 5$$

$$8x_1 + 5x_2 - 5x_3 = 25$$

An inspection of this system reveals that the third equation may be obtained by multiplying the second equation by 3 and adding both sides to the first. Thus, the third equation provides no additional information on the relationship among the variables; hence, there exist effectively only two equations in the three variables, and there exists no unique solution. Such linear relationships among the equations are known as *linear dependencies* and cause corresponding coefficient matrices to be singular. A complementary concept is that of the *rank* of a matrix, which indicates the maximum number of linearly independent relationships represented by the matrix. A matrix whose rank is less than its order is said to be of less than full rank and is, of course, singular.

1.5 A PROCEDURE FOR MATRIX INVERSION

The definition of the inverse of a matrix does not suggest a procedure for obtaining an inverse. One of the basic theorems on the existence of an inverse does suggest such a procedure (the method of cofactors) but it is quite tedious. However, there exist a large number of algorithms which specify procedures that produce matrices (normally) having the essential properties of inverses. Algorithms do not guarantee an appropriate inverse; for example, round-off errors often produce a matrix which will not have the properties of an inverse.

Regression analyses involve exclusively the inversion of *symmetric* matrices. Therefore, we present a specialized algorithm which is only useful for obtaining inverses of symmetric matrices with the aid of nonprogrammable desk calculators. The algorithm is one of several variants of the *abbreviated Doolittle* or *Gauss-Doolittle* method; the version presented here has the advantage that it provides several important answers before the entire procedure is completed.

Virtually all regression analyses are now performed on high-speed computers. Programs for performing such analyses usually do not use the abbreviated Doolittle method, since it is not the most efficient method for such computers. It is, however, useful for students to become somewhat familiar with this method, as the actual carrying out of these steps, though admittedly tedious, provides for a better understanding of the computational problems of round-off, checking, etc., as well as the sequence of steps that eventually produce answers to a regression problem.

The algorithm will be illustrated by obtaining the inverse of a 4 × 4 matrix, A; as a by-product, the procedure also provides a solution to the system of equations

$$A\underline{x} = \underline{g}$$

Generalization to other orders of matrices is not difficult.

Denote the elements of the symmetric matrix A as a_{ij} and of $\underline{g}$ as a_{ig}, that is, the g-th column of A. Write the upper right portion of this matrix on a worksheet (see Fig. 1.1), together with a check-sum column (labeled a_{ic}). This column is the sum of *all* elements of corresponding rows of A, *including* those elements which were not written down on the worksheet.

a_{11}	a_{12}	a_{13}	a_{14}	a_{1g}	a_{1c}
	a_{22}	a_{23}	a_{24}	a_{2g}	a_{2c}
		a_{33}	a_{34}	a_{3g}	a_{3c}
			a_{44}	a_{4g}	a_{4c}
a'_{11}	a'_{12}	a'_{13}	a'_{14}	a'_{1g}	a'_{1c}
b_{11}	b_{12}	b_{13}	b_{14}	b_{1g}	b_{1c}
	a'_{22}	a'_{23}	a'_{24}	a'_{2g}	a'_{2c}
	b_{22}	b_{23}	b_{24}	b_{2g}	b_{2c}
		a'_{33}	a'_{34}	a'_{3g}	a'_{3c}
		b_{33}	b_{34}	b_{3g}	b_{3c}
			a'_{44}	a'_{4g}	a'_{4c}
			b_{44}	b_{4g}	b_{4c}

Notes: (1) If $i > j$, then a'_{ij} are automatically zero; therefore these are not computed. (2) $b_{ii} = 1$ for all i. (3) In all statistical matrices, $a'_{ii} \geq 0$.

Formulas:

$$a'_{1j} = a_{1j} \qquad a'_{3j} = a_{3j} - a'_{13}b_{1j} - a'_{23}b_{2j}$$

$$b_{1j} = \frac{a'_{1j}}{a'_{11}} \qquad b_{3j} = \frac{a'_{3j}}{a'_{33}}$$

$$a'_{2j} = a_{2j} - a'_{12}b_{1j} \qquad a'_{4j} = a_{4j} - a'_{14}b_{1j} - a'_{24}b_{2j} - a'_{34}b_{3j}$$

$$b_{2j} = \frac{a'_{2j}}{a'_{22}} \qquad b_{4j} = \frac{a'_{4j}}{a'_{44}}$$

Figure 1.1 Worksheet for the abbreviated Doolittle solution (forward portion).

At this point follow the procedure outlined in Fig. 1.1. The check column is calculated in the same manner as the elements in the body of the table; the check for correct computations is that the indicated row sums must be equal (except for round-off) to the calculated elements.

The calculations outlined in Fig. 1.1 comprise the "forward" portion of the procedure, which has accomplished a transformation of the original set of equations into a new set of equations with the same solution but with a triangular coefficient matrix. This equation set, with coefficients as indicated from the worksheet, is (remember $b_{ii} = 1$)

$$\begin{aligned} x_1 + b_{12}x_2 + b_{13}x_3 + b_{14}x_4 &= b_{1g} \\ x_2 + b_{23}x_3 + b_{24}x_4 &= b_{2g} \\ x_3 + b_{34}x_4 &= b_{3g} \\ x_4 &= b_{4g} \end{aligned}$$

It is now possible to obtain the solution for values of x_i starting with x_4:

$$\begin{aligned} x_4 &= b_{4g} \\ x_3 &= b_{3g} - b_{34}x_4 \\ x_2 &= b_{2g} - b_{23}x_3 - b_{24}x_4 \\ &\cdots\cdots\cdots\cdots \end{aligned}$$

Note that if only the solution of the system of equations is needed, it is not necessary to continue further.

To obtain the inverse, which is denoted C, with elements c_{ij}, we follow the procedures given below. Note that since C must be symmetric, it is necessary to compute only the upper right portion:

$$c_{44} = \frac{1}{a'_{44}}$$

$$c_{34} = -c_{44}b_{34}$$

$$c_{24} = -c_{34}b_{23} - c_{44}b_{24}$$

$$c_{14} = -c_{24}b_{12} - c_{34}b_{13} - c_{44}b_{14}$$

These calculations have produced the last row of $C = A^{-1}$. By the definition of the inverse, the sum of products of these elements with the elements of the last row of A must be equal to 1; if not, a calculation error has occurred. Next obtain

$$c_{33} = \frac{1}{a'_{33}} - c_{34}b_{34}$$

$$c_{23} = -c_{33}b_{23} - c_{34}b_{24}$$

$$c_{13} = -c_{23}b_{12} - c_{33}b_{13} - c_{34}b_{14}$$

Again, check by multiplying by elements of the third row of A; note that c_{34} has been calculated earlier.

$$c_{22} = \frac{1}{a'_{22}} - c_{23}b_{23} - c_{24}b_{24}$$

$$c_{12} = -c_{22}b_{12} - c_{23}b_{13} - c_{24}b_{14}$$

Again, check by multiplying by elements of the second row of A.

$$c_{11} = \frac{1}{a'_{11}} - c_{12}b_{12} - c_{13}b_{13} - c_{14}b_{14}$$

Check by multiplying by elements of the first row. It is unlikely (but possible) that the above checks have allowed undetected calculation errors; hence, it is advisable to check if other than diagonal elements of CA are actually equal to zero (except for round-off) and that $\underline{x} = C\underline{g}$ agrees with the results obtained from the forward solution. Note that this part of the procedure does not require elements of the g column.

1.6 OPERATIONS WITH PARTITIONED MATRICES

It is possible to consider a matrix to be composed of several submatrices; the matrix is said to be *partitioned*. The matrices

$$P = \left[\begin{array}{ccc|c} 4 & 9 & 5 & 6 \\ 1 & 2 & 4 & 5 \\ \hline 9 & 6 & 2 & 1 \end{array}\right] \qquad R = \left[\begin{array}{cc} 4 & 2 \\ 1 & 5 \\ 2 & 2 \\ \hline 2 & 0 \end{array}\right]$$

can be partitioned as indicated by dashed lines. The submatrices are denoted by capital letters with subscripts showing location of a submatrix as if elements (unnecessary subscripts are deleted). Thus, from the above matrices,

$$P_{11} = \begin{bmatrix} 4 & 9 & 5 \\ 1 & 2 & 4 \end{bmatrix}$$

$$P_{12} = \begin{bmatrix} 6 \\ 5 \end{bmatrix}$$

$$R_2 = [2 \quad 0]$$

....

It is possible to perform all of the standard matrix operations using submatrices as if they were elements. The results of operations on partitioned matrices will be the same as if performed on the elements of the full matrices. It is, of course, necessary that all relevant submatrices be of appropriate order to permit conformability for all required operations.

The matrices P and R may thus be multiplied as follows:

$$PR = \begin{bmatrix} P_{11}R_1 + P_{12}R_2 \\ P_{21}R_1 + P_{22}R_2 \end{bmatrix}$$

Of particular interest is partitioning a square matrix into four submatrices

$$A = \begin{bmatrix} A_{11} & A_{12} \\ A_{21} & A_{22} \end{bmatrix}$$

where A_{11} and A_{22} are square, and investigating the properties of the inverse of A and applications to systems of linear equations.

Denote

$$A = \begin{bmatrix} A_{11} & A_{12} \\ A_{21} & A_{22} \end{bmatrix} \qquad C = \begin{bmatrix} C_{11} & C_{12} \\ C_{21} & C_{22} \end{bmatrix}$$

where A, C, A_{11}, A_{22}, C_{11}, and C_{22} are square; A and C, A_{11} and C_{11}, etc., are of the same order. Suppose $C = A^{-1}$, then

$$AC = I$$

or

$$\begin{bmatrix} A_{11} & A_{12} \\ A_{21} & A_{22} \end{bmatrix}\begin{bmatrix} C_{11} & C_{12} \\ C_{21} & C_{22} \end{bmatrix} = \begin{bmatrix} I & 0 \\ 0 & I \end{bmatrix}$$

Using the rules for matrix multiplication:

$$A_{11}C_{11} + A_{12}C_{21} = I$$
$$A_{11}C_{12} + A_{12}C_{22} = 0$$
$$A_{21}C_{11} + A_{22}C_{21} = 0$$
$$A_{21}C_{12} + A_{22}C_{22} = I$$

These four equations are solved for C_{11}, C_{12}, C_{21}, and C_{22} as follows:

$$\begin{aligned} C_{11} &= (A_{11} - A_{12}A_{22}^{-1}A_{21})^{-1} \\ C_{22} &= (A_{22} - A_{21}A_{11}^{-1}A_{12})^{-1} \\ C_{21} &= -A_{22}^{-1}A_{21}C_{11} \\ C_{12} &= -A_{11}^{-1}A_{12}C_{22} \end{aligned} \qquad (1.10)$$

It should be pointed out that inverting by partitioning does not result in a savings of computer time (either hand or automatic) and is thus not normally recommended as a computing procedure. The partitioning of matrices has applications to systems of linear equations, in particular, to the solution of subsets, i.e., systems of equations using subsets of equations and variables from a system.

Assume a solution has been obtained for an *entire* (m × m) system:

$$A\underline{x} = \underline{g}$$

and solution

$$\underline{x} = A^{-1}\underline{g} = C\underline{g}$$

We now want the solution to the *subset*

$$A_{11}\underline{x}_1^* = \underline{g}_1$$

in other words, to obtain A_{11}^{-1} without inverting A_{11}. Note that the solution $\underline{x}_1^*$ will *not* be equal to the first portion of $\underline{x}$, that is, $\underline{x}_1$; likewise, the inverse of A_{11} is *not* C_{11}. Using the partitioned inverse formula

$$C_{11} = (A_{11} - A_{12}A_{22}^{-1}A_{21})^{-1}$$

we can infer the converse:

$$A_{11} = (C_{11} - C_{12}C_{22}^{-1}C_{21})^{-1}$$

hence

$$A_{11}^{-1} = C_{11} - C_{12}C_{22}^{-1}C_{21} \tag{1.11}$$

an expression which involves inverting only C_{22}. This provides a considerable reduction in computing if A_{22} (hence, C_{22}) is small. Further, it can be shown that

$$\underline{x}_1^* = \underline{x}_1 - C_{12}C_{22}^{-1}\underline{x}_2 \tag{1.12}$$

i.e., the equivalent portion of the solution of the full system ($\underline{x}$) with a "correction" for the elimination of variables.

In particular, if the order of A_{22} (hence, C_{22}, $\underline{g}_2$, and $\underline{x}_2$) is 1 × 1 (a scalar), the calculations simplify considerably. Denoting $C_{22} = C_{mm}$, $\underline{g}_2 = g_m$, $\underline{x}_2 = x_m$, and elements of A_{11}^{-1} as c_{ij}^*, then

$$c_{ij}^* = c_{ij} - \frac{c_{im}c_{mj}}{c_{mm}} \tag{1.13}$$

and the elements of the solution to the subsystem are

$$x_i^* = x_i - \frac{c_{im}x_m}{c_{mm}} \tag{1.14}$$

It is, of course, not necessary for the row and column to be deleted to be the last one; the above formulas can be used with any value of m.

As an example, consider the system of equations of the form $A\underline{x} = \underline{g}$, where

$$A = \left[\begin{array}{cc|c} 2 & -1 & 0 \\ -1 & 4 & 3 \\ \hline 0 & 3 & 3 \end{array}\right] \qquad \underline{g} = \left[\begin{array}{c} 6 \\ 7 \\ \hline 9 \end{array}\right]$$

with solutions

$$C = A^{-1} = \left[\begin{array}{cc|c} 1 & 1 & -1 \\ 1 & 2 & -2 \\ \hline -1 & -2 & 7/3 \end{array}\right] \qquad \underline{x} = \left[\begin{array}{c} 4 \\ 2 \\ \hline 1 \end{array}\right]$$

It is desired to obtain the solution for x_1 and x_2 only, using the first two equations as indicated by the partitions, that is,

$$\begin{bmatrix} 2 & -1 \\ -1 & 4 \end{bmatrix} \begin{bmatrix} x_1^* \\ x_2^* \end{bmatrix} = \begin{bmatrix} 6 \\ 7 \end{bmatrix}$$

Using formula (1.13),

$$c_{11}^* = c_{11} - \frac{c_{13}c_{31}}{c_{33}}$$

$$= 1 - \frac{-1(-1)}{7/3} = \frac{4}{7}$$

The elements of the solution vector, using formula (1.14) are

$$x_1^* = x_1 - \frac{c_{13}x_3}{c_{33}} = 4 - \frac{-1(1)}{7/3} = \frac{31}{7}$$

The final results are

$$A_{11}^{-1} = C_{11}^{*} = \begin{bmatrix} 4/7 & 1/7 \\ 1/7 & 2/7 \end{bmatrix} \qquad \underline{x}_1^{*} = \begin{bmatrix} 31/7 \\ 20/7 \end{bmatrix}$$

1.7 QUADRATIC FORMS

Many derivations involving sums of squares associated with regression models result in expressions of the form

$$\underline{y}'A\underline{y}$$

where $\underline{y}$ is a vector of variables and A is a symmetric matrix of constants. This expression is known as a *quadratic form*. It can be written algebraically as

$$\sum_i \sum_j a_{ij} y_i y_j$$

or

$$\sum a_{ii} y_i^2 + 2 \sum_{i>j} a_{ij} y_i y_j$$

It is readily seen that this is a weighted sum of all possible combinations of squares and cross products of the y_i with weights given by the elements of matrix A. If A is a diagonal matrix, the quadratic form is a weighted sum of squares.

There exists a class of nonsingular symmetric matrices called *positive definite* matrices, which have the property that the value of the quadratic form is positive for *any* (non-null) vector $\underline{y}$. There are, unfortunately, no simple tests for this characteristic; one test is available as a by-product of the abbreviated Doolittle method:

> If all leading terms (a'_{ii}) of the forward solution are positive, the matrix A is positive definite

It should be noted that if A is positive definite, so is A^{-1}. As a special case, a singular matrix may be *positive semidefinite* if the value of the quadratic form is nonnegative (i.e., zero is possible) for any non-null $\underline{y}$.

1.8 PROBLEMS

1. Given the following matrices

$$A = \begin{bmatrix} 1 & 4 & 2 \\ -2 & 0 & 5 \\ 1 & -1 & 2 \end{bmatrix} \qquad B = \begin{bmatrix} 1 & 2 \\ 4 & 1 \\ 6 & 5 \end{bmatrix}$$

$$C = [5] \qquad \underline{d} = \begin{bmatrix} 2 \\ 1 \\ 1 \end{bmatrix} \qquad \underline{e} = \begin{bmatrix} 1 \\ 5 \end{bmatrix}$$

perform the following operations *if possible*; if any operation is not possible, state why.

(a) AB (b) $A\underline{e}$

(c) $C\underline{d}$ (d) $\underline{e}'C$

(e) C^{-1} (f) $\underline{d}'\underline{e}$

(g) $\underline{d}\underline{e}'$ (h) $B\underline{e}$

2. Given the set of linear equations

$$2x_1 + 2x_2 + x_3 = 4$$
$$2x_1 + 4x_2 + x_3 = 10$$
$$x_1 + x_2 + 2x_3 = 4$$

(a) Solve by forward Doolittle.

(b) Invert the coefficient matrix and obtain solution by multiplying the inverse by the righthand side.

(c) Use "reduction formulas" of Section 1.6 to obtain the solution to

$$\begin{bmatrix} 2 & 2 \\ 2 & 4 \end{bmatrix} \begin{bmatrix} x_1 \\ x_2 \end{bmatrix} = \begin{bmatrix} 4 \\ 10 \end{bmatrix}$$

that is, the system of equations containing only the first two equations and first two variables.

Chapter 2

LINEAR MODELS: ESTIMATION

2.1 DEFINITIONS AND USES OF LINEAR MODELS

The subject to be studied in this book is a subset of a broad field called *statistical data analysis*. This may include summarization of data and procedures for interpolation or extrapolation. It may also describe processes where data are used to support or deny certain relationships or hypotheses, i.e., an aid in a choice of a theoretical or conceptual model.

Statisticians have developed a wide variety of data analysis techniques, providing applications to virtually any field of study. Such techniques include not only the linear model or regression methodology described in this book but also such methods as nonparametric statistics, frequency table analysis, and a wide variety of methodologies for specialized applications. Linear model analysis is, however, the most widely used statistical data analysis technique and includes such sub-branches as the analysis of variance, the analysis of covariance, regression analysis, general linear model analysis, and extensions to multivariate analysis.

In linear model analysis the focus is on the behavior of a (usually continuous) variable such as the crop yield, the amount of chemicals obtained by a process, test scores on an examination made by students in a course, the weight gains of animals, etc. Observed data on such variables are related to certain experimental or other associated factors such as varieties or species, amounts of fertilizer or other additives, temperature, pressure, weather conditions, etc.

2.2 BASIC DEFINITIONS

The first example of statistical data analysis discussed in most elementary statistics texts is based on the *single-sample* t *test*. This test is based on the assumption of a (usually infinite) population of observable values which behave according to the linear model

$$y_i = \mu + \varepsilon_i$$

where y_i is the i-th observation from the population; μ is the mean or expected value of the y variable; and ε_i are random variables, usually independently and normally distributed, with mean zero and variance σ^2 (hence, the y_i are also random variables with mean μ and variance σ^2).

Using data from a sample of n observed values from this population, the appropriate linear model analysis provides the following familiar statistics:

$$\bar{y} = \frac{\sum y}{n}$$

is the estimate of μ,

$$s^2 = \frac{\sum (y - \bar{y})^2}{(n - 1)}$$

is the estimate of σ^2,

$$t = \frac{\bar{y} - \mu_0}{\sqrt{s^2/n}}$$

provides a test of the hypothesis H_0: $\mu = \mu_0$, and

$$\bar{y} \pm t_{\alpha/2}\sqrt{\frac{s^2}{n}}$$

is a two sided $(1 - \alpha)$-level confidence interval for μ.

A more general and frequently used linear model, usually called the *multiple linear regression model*, is stated as follows:

$$y_i = \beta_0 + \beta_1 x_{i1} + \beta_2 x_{i2} + \cdots + \beta_m x_{im} + \varepsilon_i \tag{2.1}$$

where y_i and ε_i are defined as above, x_{ij} is the observed value of the j-th of m other variables associated with the i-th y value, and β_j is a *parameter* which specifies how y is related to the x_j.

This is obviously a more complicated model than the one given for the single sample t test, as it involves a number of different variables. The variable represented by y is the variable of interest, that is, the variable to be analyzed or studied. The variables represented by the x_j are associated with y and may, in some manner, influence the behavior of the y variable. The y variable is thus known as the *dependent* variable and the x variables are known as the *independent* variables, and the regression model is often referred to as the regression *of* y *on* the x variables. The partial regression coefficients β specify the linear functional relationship between the independent variables and the dependent variable; specifically, β_j indicates the change in the dependent variable (y) associated with a unit change in the corresponding independent variable (x_j), all other independent variables remaining constant. Mathematically, the β_j are the partial derivatives of the functional relationship $\partial y/\partial x_j$. Consequently, they are known as *partial regression coefficients* (see Section 2.5). The parameter β_0 is known as the y-*intercept* and represents the expected value of y when all x are zero; the value of this parameter may not have any practical meaning, as in many applications, it may lie outside the range of relevant data.

Applications of this model are numerous; a perusal of journals in the applied sciences, especially in the biological, agricultural, social, and behavioral sciences, will reveal a large number of such examples. A few examples of typical applications may be useful.

1. The relationship of the dependent variable crop yield to the independent variables x_1, x_2, x_3, which are amounts of fertilizer components, (N, P, K). The regression coefficients estimate the gain in yield resulting from the application of one unit of each of these components, holding amounts of other components constant.

2. The relationship of the yield of a chemical process to pressure, temperature, and the introduction of various additives or catalysts.
3. The consumption of a consumer good, say pork, as related to the price of pork and prices of competing goods, such as beef, chicken, or turkey, as well as income.
4. A production function relating the physical output of a production plant or process to various inputs such as labor, capital, quantities of physical inputs, power consumption, etc.
5. The performance of college students, as measured by their grade point average, related to their high school performance, size of high school, and performance on various entrance and/or aptitude tests.

An alternate formulation of the model is

$$y_i = \mu_x + \varepsilon_i \tag{2.2a}$$

and

$$\mu_x = \beta_0 + \beta_1 x_{i1} + \beta_2 x_{i2} + \cdots + \beta_m x_{im} \tag{2.2b}$$

Note that model statement (2.2a) looks very much like the model of the t test discussed above, except that the mean μ_x, instead of being single-valued, takes on a unique value for any specific set of values of the x variables. This is then the *conditional mean* of the population of y-values for a specific set of x-values.

It is a widely accepted custom to simplify notation by eliminating the subscript i denoting the i-th observation; hence, the model is usually written

$$y = \beta_0 + \beta_1 x_1 + \beta_2 x_2 + \cdots + \beta_m x_m + \varepsilon \tag{2.3}$$

There are two algebraically equivalent model statements which will occasionally be useful. The first is

$$y = \beta_0 x_0 + \beta_1 x_1 + \beta_2 x_2 + \cdots + \beta_m x_m + \varepsilon \tag{2.4}$$

where x_0 is a "dummy" variable which is identically equal to unity. The second is

$$y = \mu + \beta_1(x_1 - \bar{x}_1) + \beta_2(x_2 - \bar{x}_2) + \cdots + \beta_m(x_m - \bar{x}_m) + \varepsilon \quad (2.5)$$

where μ is the (unconditional) mean of y, and $\bar{x}_1, \bar{x}_2, \ldots, \bar{x}_m$ are sample means of the $x_1, x_2, \ldots, x_m$, respectively. It follows that (2.5) is equivalent to (2.3) by defining

$$\beta_0 = \mu - \beta_1\bar{x}_1 - \beta_2\bar{x}_2 - \cdots - \beta_m\bar{x}_m \quad (2.6)$$

A regression *analysis* consists of using sample data to estimate the β coefficients and σ^2 and to make various tests of hypotheses or confidence interval statements about the parameters. This is usually accomplished by a method called *least squares* (Section 2.4), which results in an estimating equation for μ_x:

$$\hat{\mu}_x = \hat{\beta}_0 + \hat{\beta}_1x_1 + \hat{\beta}_2x_2 + \cdots + \hat{\beta}_mx_m \quad (2.7)$$

where $\hat{\mu}_x$ is the estimate for μ_x, $\hat{\beta}_j$ are estimates of β_j, and σ^2 is estimated as a by-product of the least-squares procedure (Chapter 3).

2.3 ASSUMPTIONS AND LIMITATIONS

The proper use of a linear regression model to analyze a set of data is subject to a number of assumptions and limitations. Failure of these assumptions and limitations may require the modification of the model or the placing of restrictions on the inference to be made by the analyses. Occasionally it may even lead to the abandonment of a regression analysis. It is therefore appropriate to examine these assumptions and limitations before proceeding to the estimation and inference procedures.

1. The most important condition is, of course, that the model has been *correctly specified*, i.e., the model adequately and correctly describes the behavior of the data. It is important to keep in mind, however, that the model need not

correctly describe the actual physical phenomenon underlying the data, but only that the model adequately describes the behavior of the data. Further discussion on this topic is provided in Chapter 4.

2. The linear regression model must be linear in the *parameters*. However, the model may be nonlinear in the independent *variables* and yet remain within the framework of linear regression. This is discussed further in Section 2.8 and again in Chapter 6.
3. The model statement makes no mention of *statistical properties* of the *independent variables*. This is intentional, since they are not random variables. They are specified to be fixed, predetermined, and/or chosen at will. They may, for example, be the settings of temperature and pressure controls in an experiment to relate the yield of a chemical process to temperature and pressure. Independent variables may take any form: frequencies, measured quantities, dummy indicator variables such as a "one" to indicate the presence of a condition and a "zero" for absence, etc. There are, however, many practical situations where the independent variables may not be strictly regarded as being fixed. For example, the data may consist of temperatures and associated yields observed during the actual operation of a chemical process. A regression analysis is, however, usually valid for such situations; the important condition is that no inferences are to be made about the independent variables and that the independent variables are not influenced by the dependent variable. The focus of *all* inferences is on the dependent variable.
4. The independent variables must be *measured without error*. For example, the recorded temperature must be the true temperature. If the independent variables are subject to error, the estimated regression coefficients may be biased. This bias is a function of the ratio of the measurement

error to the random error of the model; if this ratio is small, the bias is small.

5. The stated purpose of the regression model and analysis per se is to estimate μ_x, the mean of the dependent variable for a specific set of values of the independent variables. Normally, such estimation is performed within the range of the observed sample data and extrapolation is not encouraged (see Section 6.2 for an example).
6. Another purpose for performing regression analyses is to study the partial regression coefficients. Under certain conditions, the regression coefficient estimates are quite unstable and not subject to meaningful interpretation. This is discussed in Chapter 5.
7. The fact that a regression relationship has been found to exist does *not* imply a cause-effect relationship, although it may exist. For example, the smoking-lung cancer controversy hinges largely on the question of whether the demonstrated relationship between smoking and lung cancer does indeed imply that smoking *causes* lung cancer. A counterargument is that some outside agent may cause this relationship to exist.
8. The data on which the regression analysis is performed is assumed to be a random sample. However, regression analysis is often used in situations where this is not strictly the case. This includes the analysis of time series or data where all members of a population may have been observed, such as all the students in a particular class. In such cases, regression analysis is still a valuable tool for the description of the data and possible inferences to similar conditions.

2.4 ESTIMATION OF PARAMETERS BY LEAST SQUARES

Having specified a model, it is appropriate to develop procedures which use observed data to provide estimates of the parameters, to test hypotheses, and for other aspects of statistical inference. Procedures for estimation of parameters are given in this chapter, while other aspects of statistical inference are discussed in Chapter 3.

The estimating principle to be used is that of *least squares*. There exist a number of other principles of estimation which are discussed in texts on statistical theory. However, the least-squares procedure enjoys the widest use by virtue of its simplicity and other favorable properties.

Consider a set of data on n observations of y and associated x variables which has been hypothesized to behave according to a regression model as specified above:

$$y_i = \beta_0 + \beta_1 x_{i1} + \beta_2 x_{i2} + \cdots + \beta_m x_{im} + \varepsilon_i \tag{2.1}$$

or

$$y_i = \mu_x + \varepsilon_i \qquad (i = 1, 2, \ldots, n) \tag{2.2}$$

It is required to obtain estimates of the β_j, denoted $\hat{\beta}_j$, which provide estimates of μ_x:

$$\hat{\mu}_x = \hat{\beta}_0 + \hat{\beta}_1 x_{i1} + \hat{\beta}_2 x_{i2} + \cdots + \hat{\beta}_m x_{im} \tag{2.7}$$

The least-squares principle specifies that the $\hat{\beta}_j$ are to be chosen so as to minimize the sum of squared differences between the actually observed values and the estimated values of the dependent variable, that is, the y_i and $\hat{\mu}_x$, respectively. This quantity is known as the *residual* or *error sum of squares*:

$$SSE = \sum_{i=1}^{n} (y_i - \hat{\mu}_x)^2 = \sum(y_i - \hat{\beta}_0 - \hat{\beta}_i x_{i1} - \hat{\beta}_2 x_{i2} - \cdots - \hat{\beta}_m x_{im})^2 \quad (2.8)$$

The minimization of this quantity, which results in the least-squares estimates, is accomplished by a relatively simple exercise in differential calculus. However, since calculus is not a prerequisite to the understanding of this book, it is instructive to demonstrate the least-squares principle by an example. The first two columns of Table 2.1 give a set of five pairs of observations of x and y which are to be used to estimate the coefficients of the one-variable regression model:

$$y = \beta_0 + \beta_1 x + \varepsilon \quad (2.9)$$

TABLE 2.1 Example for Least-Squares Estimation

		Estimating equations					
		1	2	3	4	5	6
		$\hat{\beta}_0 = 1.0$	$\hat{\beta}_0 = 0.0$	$\hat{\beta}_0 = 2.0$	$\hat{\beta}_0 = 1.0$	$\hat{\beta}_0 = 1.0$	$\hat{\beta}_0 = 2.2$
		$\hat{\beta}_1 = 2.0$	$\hat{\beta}_1 = 2.0$	$\hat{\beta}_1 = 2.0$	$\hat{\beta}_1 = 1.5$	$\hat{\beta}_1 = 2.5$	$\hat{\beta}_1 = 1.6$
Actual observations		Estimated values					
x	y	$\hat{\mu}_x$	$\hat{\mu}_x$	$\hat{\mu}_x$	$\hat{\mu}_x$	$\hat{\mu}_x$	$\hat{\mu}_x$
1	3	3.0	2.0	4.0	2.5	3.5	3.8
2	7	5.0	4.0	6.0	4.0	6.0	5.4
3	7	7.0	6.0	8.0	5.5	8.5	7.0
4	7	9.0	8.0	10.0	7.0	11.0	8.6
5	11	11.0	10.0	12.0	8.5	13.5	10.2
$\Sigma(y - \hat{\mu}_x)^2$		8.0	13.0	13.0	17.75	25.75	6.4

Figure 2.1 shows the five data points as (⊙). The estimation problem is to determine the parameters (β_0, β_1) of the straight line

$\hat{\mu}_x = \beta_0 + \beta_1 x$ which "best" describes the plotted points. One candidate for such a line, indicated by the dashed line (in Figure 2.1), is

$$\hat{\mu}_x = 1.0 + 2.0x$$

that is, $\hat{\beta}_0 = 1$ and $\hat{\beta}_1 = 2$. This equation appears to fit the data rather well, and it is easy to verify (under equation 1 of Table 2.1) that SSE = 8.0. Four other "reasonable" equations produce the values of SSE given in Table 2.1. If we choose only from this set, the first equation would seem to be the least-squares equation. The "correct" values, produced by the least-squares procedure (below), obtain $\hat{\beta}_0 = 2.2$ and $\hat{\beta}_1 = 1.6$ with a residual sum of squares of 6.4 (last column); hence, although the "hunt and inspect" method came close, it did not find the least-squares estimate.

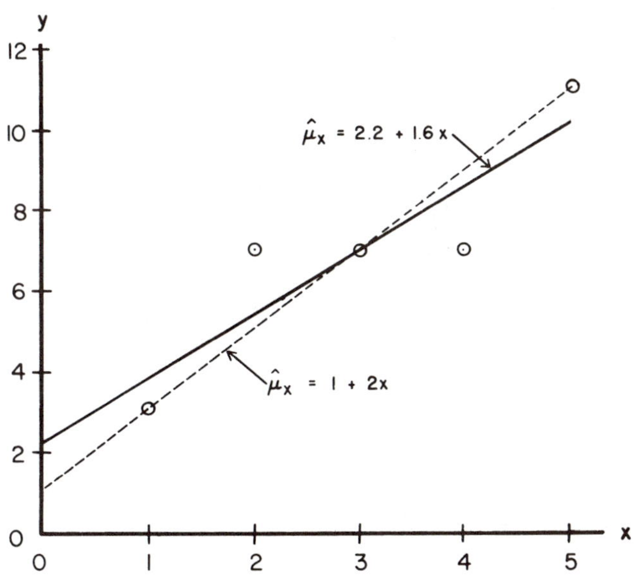

Figure 2.1 Linear regression example.

The procedure is developed by the use of differential calculus. The residual sum of squares for the model (2.9) is

$$\text{SSE} = \sum_{i=1}^{n} (y_i - \beta_0 - \beta_1 x_{i1})^2$$

The minimum SSE occurs when the partial derivatives with respect to β_0 and β_1 are zero. The partial derivatives with respect to β_0 and β_1 are

$$\begin{aligned} &\text{Partial derivative} \\ &\quad \text{w.r.t. } \beta_0 = \frac{\partial \text{SSE}}{\partial \beta_0} = -2 \sum_{i=1}^{n} (y_i - \beta_0 - \beta_1 x_{i1}) \\ &\text{Partial derivative} \\ &\quad \text{w.r.t. } \beta_1 = \frac{\partial \text{SSE}}{\partial \beta_1} = -2 \sum_{i=1}^{n} x_{i1}(y_i - \beta_0 - \beta_1 x_{i1}) \end{aligned} \tag{2.10}$$

Setting both of these equal to zero, dividing each equation by -2, and rearranging terms produces the following two equations, usually called the *normal equations* (the obvious subscripts are dropped):

$$\begin{aligned} \beta_0 n + \beta_1 \sum x &= \sum y \\ \beta_0 \sum x + \beta_1 \sum x^2 &= \sum xy \end{aligned} \tag{2.11}$$

The solution for β_0 and β_1 which satisfies these equations provides the desired estimates. Rearranging the first equation results in

$$\hat{\beta}_0 = \bar{y} - \hat{\beta}_1 \bar{x} \tag{2.12}$$

Substituting into the second equation and rearranging produces the estimate

$$\hat{\beta}_1 = \frac{\sum xy - \sum x \sum y/n}{\sum x^2 - (\sum x)^2/n} \tag{2.13}$$

Most readers will recognize these formulas from their first course in statistics as that version of the estimates most suitable for manual computation. Equation (2.13) is algebraically equivalent to

$$\hat{\beta}_1 = \frac{\sum(x - \bar{x})(y - \bar{y})}{\sum(x - \bar{x})^2} \tag{2.14}$$

which reveals more clearly the structure of the estimate as the means corrected (centered) sum of cross products of x and y divided by the sum of squares of x-deviations from their sample mean. Substituting the values in the example produces the estimates $\hat{\beta}_0 = 2.2$ and $\hat{\beta}_1 = 1.6$. As in the case of the unconditional mean, the sum of residuals from the conditional mean estimates $\Sigma(y - \hat{\mu}_x)$ is zero.

Estimates of parameters of the multiple regression model

$$y = \beta_0 + \beta_1 x_1 + \beta_2 x_2 + \cdots + \beta_m x_m + \varepsilon \tag{2.3}$$

are obtained by applying the same principle of obtaining the partial derivatives of SSE with respect to $\beta_0, \beta_1, \ldots, \beta_m$ and setting each equal to zero. After rearranging terms, a set of normal equations is obtained:

$$\begin{aligned}
\beta_0 n + \beta_1 \sum x_1 + \beta_2 \sum x_2 + \cdots + \beta_m \sum x_m &= \sum y \\
\beta_0 \sum x_1 + \beta_1 \sum x_1^2 + \beta_2 \sum x_1 x_2 + \cdots + \beta_m \sum x_1 x_m &= \sum x_1 y \\
\beta_0 \sum x_2 + \beta_1 \sum x_2 x_1 + \beta_2 \sum x_2^2 + \cdots + \beta_m \sum x_2 x_m &= \sum x_2 y \\
\cdots\cdots\cdots\cdots\cdots\cdots\cdots\cdots\cdots\cdots\cdots\cdots \\
\beta_0 \sum x_m + \beta_1 \sum x_m x_1 + \beta_2 \sum x_m x_2 + \cdots + \beta_m \sum x_m^2 &= \sum x_m y
\end{aligned} \tag{2.15}$$

Note that the use of only the first two equations and terms involving only β_0 and β_1 reproduces the normal equations for the one-variable regression (2.11). Because of the large number of equations, it is not possible to obtain a convenient expression for the estimates of the β coefficients from this set of equations: the β coefficients are obtained by solving the above set of m + 1 simultaneous linear equations in m + 1 variables.

An alternate formulation can be obtained by solving the first equation in (2.15) for β_0:

$$\beta_0 = \bar{y} - \beta_1 \bar{x}_1 - \beta_2 \bar{x}_2 - \cdots - \beta_m \bar{x}_m \tag{2.16}$$

Substituting this expression for β_0 in the second equation of (2.15) and rearranging produces

$$\beta_1(\sum x_1^2 - n\bar{x}_1^2) + \beta_2(\sum x_1x_2 - n\bar{x}_1\bar{x}_2) + \cdots + \beta_m(\sum x_1x_m - n\bar{x}_1\bar{x}_m) = \sum x_1y - n\bar{x}_1\bar{y} \qquad (2.17)$$

with similar expression for the other equations. Recalling that for any two variables, say x and y,

$$\sum(x - \bar{x})(y - \bar{y}) = \sum xy - \frac{\sum x \sum y}{n} = \sum xy - n\bar{x}\bar{y}$$

it can be seen that in equation (2.17) the coefficients of the β_j are in terms of the sums of squares and cross products of the variables *corrected for their means* and the number of equations to be solved has been reduced to m. For this reason and also due to better control of round-off error, this version is widely used for the numerical solution of the normal equations.

The use of the dummy variable in the model equation (2.4) produces normal equations which are identical with (2.15). The use of model equation (2.5) produces directly the normal equations (2.17) with the additional estimate

$$\hat{\mu} = \bar{y}$$

A more compact presentation of the normal equations is obtained using matrix notation. Define

$$\underline{y} = \begin{bmatrix} y_1 \\ y_2 \\ \vdots \\ y_n \end{bmatrix}$$

the column vector of the n observed values of y ($\underline{\hat{\mu}}_x$ is the corresponding vector of the estimated conditional means),

$$X = \begin{bmatrix} 1 & x_{11} & x_{12} & \cdots & x_{1m} \\ 1 & x_{21} & x_{22} & \cdots & x_{2m} \\ \cdots & \cdots & \cdots & \cdots & \cdots \\ 1 & x_{n1} & x_{n2} & \cdots & x_{nm} \end{bmatrix}$$

the n × (m + 1) matrix of n observations on the m independent (x) variables plus the dummy variable, and

$$\underline{\beta} = \begin{bmatrix} \beta_0 \\ \beta_1 \\ \beta_2 \\ \vdots \\ \beta_m \end{bmatrix}$$

the column vector of the m + 1 partial regression coefficients ($\hat{\underline{\beta}}$ is the vector of estimated regression coefficients).

The regression model using these matrices is written

$$\underline{y} = X\underline{\beta} + \underline{\varepsilon} \tag{2.1a}$$

where $\underline{\varepsilon}$ is an n-valued vector of the ε_i.

The expression

$$\hat{\underline{\mu}}_x = X\hat{\underline{\beta}} \tag{2.18}$$

gives the estimated regression model, equivalent to (2.7), and

$$\underline{y} - \hat{\underline{\mu}}_x \tag{2.19}$$

is the vector of residuals. The residual sum of squares (2.8) is obtained by the operation

$$(\underline{y} - \hat{\underline{\mu}}_x)'(\underline{y} - \hat{\underline{\mu}}_x) \tag{2.20}$$

which, when minimized with respect to elements of $\hat{\underline{\beta}}$, produces the system of equations represented by

$$(X'X)\underline{\beta} = X'\underline{y} \tag{2.21}$$

which is the matrix equivalent of the normal equations (2.11). The matrix X'X is a symmetric matrix consisting of sums of squares and cross products of the x variables; the $X'\underline{y}$-vector is a vector of the xy cross products. The system is solved to obtain the estimate

$$\hat{\underline{\beta}} = (X'X)^{-1}X'\underline{y} \tag{2.22}$$

A frequently used convention is to denote

$$C = (X'X)^{-1}$$

$$\underline{g} = X'\underline{y}$$

hence,

$$\hat{\underline{\beta}} = C\underline{g} \tag{2.23}$$

The solution $\hat{\underline{\beta}}$ is obtained by inverting X'X and then multiplying the inverse by the vector $X'\underline{y}$ as outlined in Chapter 1.

The vector $\hat{\underline{\beta}}$ can be obtained by solving only m equations by using sums of squares and cross products of *deviations* from means, i.e., corrected for the mean [see equations (2.5) and (2.17)]. In this case, define

$$X = \begin{bmatrix} (x_{11} - \bar{x})(x_{12} - \bar{x}_2) & \cdots & (x_{1m} - \bar{x}_m) \\ (x_{21} - \bar{x}_1)(x_{22} - \bar{x}_2) & \cdots & (x_{2m} - \bar{x}_m) \\ \cdots\cdots\cdots\cdots & \cdots & \cdots\cdots \\ (x_{n1} - \bar{x}_1)(x_{n2} - \bar{x}_2) & \cdots & (x_{nm} - \bar{x}_m) \end{bmatrix} \tag{2.24}$$

and

$$\underline{y} = \begin{bmatrix} y_1 - \bar{y} \\ y_2 - \bar{y} \\ \vdots \\ y_n - \bar{y} \end{bmatrix}$$

and now

$$\hat{\underline{\beta}} = (X'X)^{-1}X'\underline{y} \tag{2.22a}$$

which is the same expression as (2.22) except that X'X and X'$\underline{y}$ are defined to be the matrices of sums of squares and cross products corrected for means and are usually called the *corrected sum of squares* and *cross product matrices*. The estimate $\hat{\beta}_0$ is calculated separately by the already developed formula (2.16):

$$\hat{\beta}_0 = \bar{y} - \hat{\beta}_1\bar{x}_1 - \hat{\beta}_2\bar{x}_2 - \cdots - \hat{\beta}_m\bar{x}_m \tag{2.16}$$

2.5 THE PARTIAL REGRESSION COEFFICIENT

The vector of regression coefficients, $\hat{\underline{\beta}}$, comprises the set of *partial* regression coefficients as defined in Section 2.2. This is in contrast to the set of *total* regression coefficients which would be obtained if separate linear regressions would be performed using individually each independent variable. The differences in interpretation and usage of the partial and total regression coefficients are often misunderstood. In this section these differences are explained in three ways:

1. A simple conceptual example
2. A numerical example
3. An alternate method of obtaining a partial regression coefficient estimate

1. Consider the problem of determining the relationship of students' knowledge of statistics to intelligence and the number of statistics courses taken. Let y be the test score used to determine knowledge of statistics, x_1 the intelligence quotient (IQ) score, and x_2 the number of statistics courses taken. One would *not* expect very useful results from the model

$$y = \beta^*_{01} + \beta^*_1 x_1 + \epsilon^*_1$$

(the asterisks denoting that these are total regression coefficients), since IQ *by itself* would not cause a person to do well on a statistics test. One would expect somewhat better results using the model:

$$y = \beta^*_{02} + \beta^*_2 x_2 + \varepsilon^*_2$$

since exposure to statistics courses should be related to a student's knowledge. The multiple regression, using the model equation

$$y = \beta_0 + \beta_1 x_1 + \beta_2 x_2 + \varepsilon$$

should, however, show *both* x_1 and x_2 to be useful in estimating test scores, since the partial coefficient β_1 measures the effect of IQ on performance for students having *equal* exposure to statistics courses, while the partial coefficient β_2 indicates effect of additional courses for students with *equal* IQ. In other words, the *total* regression coefficients β^*_1 and β^*_2 are not particularly useful, whereas the partial coefficients β_1 and β_2 are. This is illustrated in Figure 2.2 which shows how IQ will tend to improve the test score for any level of exposure in classes (although the number of courses is most important).

2. It is, in fact, quite possible that the total and partial regression coefficient estimates are completely different. This is illustrated below by an artificial example (the data are given in Table 2.2); however, similar conditions are often found with "real" data.

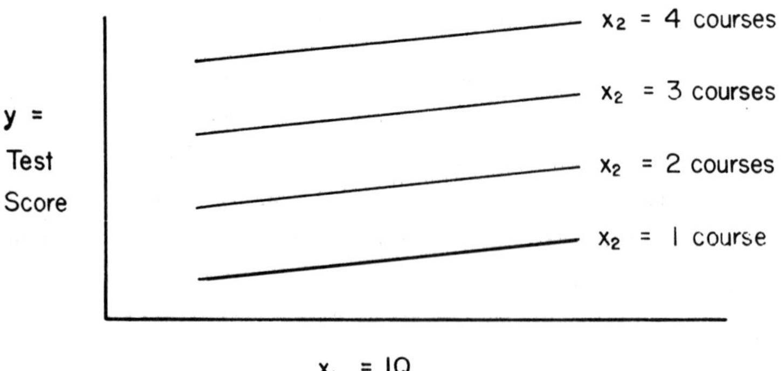

Figure 2.2 Hypothetical relationship between IQ and course exposure and test-scores.

TABLE 2.2 Artificial Data

Obs. No.	x_1	x_2	y
1	0	2	2
2	2	6	3
3	2	7	2
4	2	5	7
5	4	9	6
6	4	8	8
7	4	7	10
8	6	10	7
9	6	11	8
10	6	9	12
11	8	15	11
12	8	13	14

Estimating the parameters of the total regression of y using only x_1 produces the equation

$$\hat{\mu}_{x_1} = 1.86 + 1.30x_1$$

while using only x_2 produces

$$\hat{\mu}_{x_2} = 0.86 + 0.78x_2$$

The estimated *multiple* regression equation is

$$\hat{\mu}_x = 5.37 + 3.01x_1 - 1.29x_2$$

The differences between the two sets of coefficients stand out clearly: the estimated total coefficients are 1.30 and 0.78 for x_1 and x_2, respectively, while the corresponding partial coefficients are 3.01 and -1.29. The reason for the difference between the total and partial coefficients for x is illustrated in Figure 2.3. *Ignoring* the x_1 values, the scattering of all the points indicates a positively sloping relationship between y and x_2. However, an examination of values of y and x for constant values of x_1 (say $x_1 = 4$) shows a relationship between y and x_2 with a negative slope.

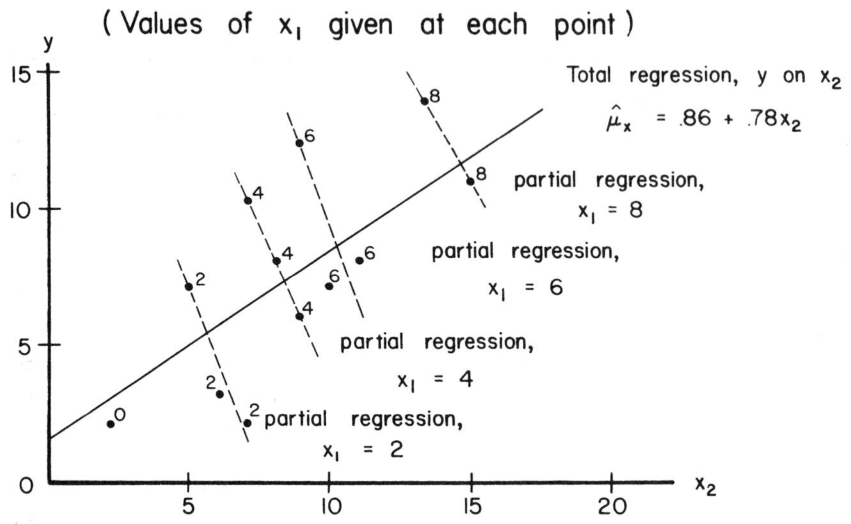

Figure 2.3 Relationships between x_2 and y in the artificial example (values of x_1 given at each point).

The partial regression $\beta_2 = -1.29$ is an average of the negative relationships between y and x_2 for specific values of x_1, as indicated in Figure 2.3.

3. A third way of illustrating the meaning of a partial regression coefficient can be obtained by the use of residuals for estimating, say, β_2 of a two-variable regression model.

Perform a *regression* using as the *independent* variable x_1 and as the *dependent* variable x_2, using the model $x_2 = \gamma_0 + \gamma_1 x_1 + \varepsilon$. Use the estimated coefficients ($\hat{\gamma}_0$, $\hat{\gamma}_1$) to obtain estimated values of x_2, denoted $\hat{x}_2$, and residuals $\hat{d}_2 = (x_2 - \hat{x}_2)$ for each observed value of x_2. These *residuals* are those parts of the x_2 observations which *cannot* be explained by (or corrected for) a linear relationship with x_1. Now perform a (total) linear regression of y on $\hat{d}_2$; the resulting coefficient is the least-squares estimate of the *partial* regression coefficient β_2 of the model involving both x_1 and x_2. It is important to note that this result is *not* obtained if, instead, the *dependent* variable y is "corrected" for x_1. In other words, the partial regression coefficient is the effect due to the corresponding *independent* variable corrected for the influence of *other independent* variables.

Clearly, then, if a multiple regression model has been proposed to describe a set of data, it may not be useful and, in fact, may be misleading to investigate the corresponding total regressions involving the individual independent variables. Thus, although (for the sake of practice and instruction) some of the problems at the end of this and the next chapter request the estimation of both total and partial coefficients, the usual procedure is to estimate only the partial regression coefficients.

2.6 AN EXAMPLE

It is desired to relate the annual cost of operating livestock auction markets to the quantity and mixture of livestock sold through the respective markets. The purpose of such a study is not only to estimate the cost of operating a livestock market for given volume and mixture but also to study the structure of costs as they relate to these factors, as this cost structure may be used as a basis for charges to be assessed sellers of livestock. For this study, cost and volume data have been assembled for 19 auction markets for a calendar year. These data are presented in Table 2.3. Note that the data have been coded to units of thousands; this makes the magnitudes of numbers more manageable but does not affect the results.

The relationship between cost and volume data is hypothesized to be of the form

$$y = \beta_0 + \beta_1 x_1 + \beta_2 x_2 + \beta_3 x_3 + \beta_4 x_4 + \varepsilon$$

where

y = annual cost of operating an auction market

x_1 = number of cattle sold

x_2 = number of calves sold

x_3 = number of hogs sold

x_4 = number of sheep sold

ε = residual (assumed to be a random variable)

TABLE 2.3 Auction Market Costs

	No. heads sold ($\times 10^3$)				
Market	Cattle	Calves	Hogs	Sheep	Cost ($\times$ \$$10^3$)
1	3.437	5.791	3.268	10.649	27.698
2	12.801	4.558	5.751	14.375	57.634
3	6.136	6.223	15.175	2.811	47.172
4	11.685	3.212	0.639	0.694	49.295
5	5.733	3.220	0.534	2.052	24.115
6	3.021	4.348	0.839	2.356	33.612
7	1.689	0.634	0.318	2.209	9.512
8	2.339	1.895	0.610	0.605	14.755
9	1.025	0.834	0.734	2.825	10.570
10	2.936	1.419	0.331	0.231	15.394
11	5.049	4.195	1.589	1.957	27.843
12	1.693	3.602	0.837	1.582	17.717
13	1.187	2.679	0.459	18.837	20.253
14	9.730	3.951	3.780	0.524	37.465
15	14.325	4.300	10.781	36.863	101.334
16	7.737	9.043	1.394	1.524	47.427
17	7.538	4.538	2.565	5.109	35.944
18	10.211	4.994	3.081	3.681	45.945
19	8.697	3.005	1.378	3.338	46.890

The first step is to obtain the matrix of corrected sums of squares and cross products (X'X), which is given in Table 2.4. Also given in this table is the vector of corrected sums of cross products between the independent variables and the dependent variable ($X'\underline{y}$). These matrices are a representation of the coefficients and right-hand side of the system of normal equations, the first of which, for example, is

$$315.995(\beta_1) + 61.922(\beta_2) + 136.656(\beta_3) + 267.706(\beta_4) = 1442.187$$

Usually, however, the equations are not specifically written in this manner; the matrices as given in Table 2.4 are adequate for practical purposes. The solution of this system of equations is usually obtained by computers, and the results are given in Table 2.5, which contains the inverse matrix and the solution vector. It will be seen in Chapter 3 that the elements of the inverse matrix are required for computing statistics for hypothesis tests, etc.

The intercept term β_0 is estimated separately according to formula (2.16); the estimated regression equation is

$$\hat{\mu}_x = 2.28843 + 3.2155x_1 + 1.6132x_2 + 0.8149x_3 + 0.8026x_4$$

TABLE 2.4 Corrected Sums of Squares and Products[a]

Variable	x_1	x_2	x_3	x_4	y
x_1	315.995	61.922	136.656	267.706	1442.187
x_2	61.922	71.324	59.283	31.460	387.724
x_3	136.656	59.283	276.590	291.198	994.148
x_4	267.706	31.460	291.198	1462.073	2322.290

[a]Data rounded for presentation.

TABLE 2.5 Inverse and Coefficients[a]

Variable	x_1	x_2	x_3	x_4	Coeff.
x_1	0.00469830	-0.00292272	-0.00108234	-0.00058180	3.21549
x_2	-0.00292272	0.01912274	-0.00352373	0.00082549	1.61316
x_3	-0.00108234	-0.00352373	0.00584200	-0.00088954	0.81488
x_4	-0.00058180	0.00082549	-0.00088954	0.00094989	0.80259

[a]Data rounded for presentation.

This equation can be used to estimate the cost for any specific market size and mixture. The first market, for example, has an estimated cost of \$33,891 (compared to the actual value of \$27,698). The difference between the two is the estimate of the residual component of the model, and it will be seen in Chapter 4 that a study of such residuals is often quite useful.

The estimated coefficients are also of interest. The intercept estimate represents the hypothetical cost of an auction market selling no animals. This could be considered an estimate of the fixed cost, but the value is an extrapolation beyond the range of the data and should not be taken seriously. The other coefficients indicate the effect on the cost of operating such a market associated with a one-animal increase in sales for each type of animal, respectively. Thus, selling one additional head of cattle is estimated to add \$3.22 to the cost of operating such a market (actually \$3216 per 1000 head), while the addition of one hog adds \$0.81. These estimates appear reasonable, in view of the physical size of the animals being sold.

2.7 STANDARDIZED REGRESSION COEFFICIENTS

The regression coefficients of the linear model are functions of the units of measurement of the independent and dependent variables. In the previous example, the coefficient for cattle (3.22) is the additional \$1000 cost of operating an auction market per 1000 *head* of cattle sold. If the number of cattle sold is expressed in simple units, the coefficient is 0.00322. Thus, the magnitudes of coefficients are influenced by choices of units of measurement. This may make interpretation of regression coefficients somewhat difficult, particularly when units of measurement of different independent variables in a model vary greatly in size.

Estimates of regression coefficients independent of the units of measurement are obtained by the use of *standardized* regression coefficients. Standardized coefficients result when the regression is based on variables which have been transformed to equal (usually unit) variance. Such standardized variables may be obtained by the equation for a specific variable x_j:

$$z_{ij} = \frac{x_{ij} - \bar{x}_j}{s_j} \qquad (i = 1, 2, \ldots, n) \tag{2.25}$$

The multiple regression equation with standardized coefficients is obtained by considering the model

$$z_y = \beta_1' z_1 + \beta_2' z_2 + \cdots + \beta_m' z_m \tag{2.26}$$

where z_y is the standardized dependent variable, and the β_i', $i = 1, 2, \ldots, m$, the standardized partial regression coefficients. Note that the intercept is zero. Each coefficient indicates how many *standard deviation* changes in y are associated with one *standard deviation* change in that x, all other x_j constant. It is easy to see that the relative magnitudes of standardized coefficients may have some meaning. The relationship between ordinary and standardized coefficients is

$$\beta_i' = \beta_i \frac{s_{x_i}}{s_y} \tag{2.27}$$

Normally, if standardized coefficient estimates are desired, the usual coefficients are first estimated and the standardized coefficients are then obtained using equation (2.27). Alternately, standardized regression coefficients may be obtained by transforming the sum of squares and cross product matrices to matrices of correlations and by using elements of these matrices as the coefficients of the normal equations.

2.8 REGRESSION THROUGH AN ARBITRARY POINT

One feature of a least-squares linear regression is that the locus of estimated values, $\hat{\mu}_x$, often called the *estimated response plane*, must pass through the origin as defined by the variables. In the usual case, all variables have been corrected for their means; hence, the origin is the point $(\bar{x}_1, \bar{x}_2, \ldots, \bar{x}_m, \bar{y})$, and the regression plane will pass through this point.

This property of the least-squares estimate can be utilized to obtain an estimated regression equation which passes through any arbitrary point. Let $(x_{10}, x_{20}, \ldots, x_{m0}, y_0)$ be such a point. Redefine a new set of variables:

$$\begin{aligned}
w_1 &= x_1 - x_{10} \\
w_2 &= x_2 - x_{20} \\
&\cdots\cdots\cdots\cdots \\
w_m &= x_m - x_{m0} \\
w_y &= y - y_0
\end{aligned}$$

Then the regression model

$$w_y = \beta_1 w_1 + \beta_2 w_2 + \cdots + \beta_m w_m + \varepsilon \qquad (2.28)$$

(note the absence of an intercept term) will produce a response plane which passes through the desired point. It should be noted, however, that a regression equation produced in this manner will *not* have the usual property

$$\sum^{n}(y - \hat{\mu}_x) = 0$$

A special use of this principle which has rather frequent application is to force the regression plane through the origin, making $\hat{\mu}_x = 0$ when all $x_i = 0$. This is accomplished by letting all x_{i0} and y_0 (above) be zero, that is, using the uncorrected (*raw*) sum of squares and cross products for X'X and X'$\underline{y}$.

2.9 CURVILINEAR REGRESSIONS

It was noted earlier that the linear regression model specifies linearity in the parameters, i.e., in the β coefficients and ε. There is, however, no similar restriction regarding linearity of the independent variables. It is therefore possible to specify such apparently nonlinear models as

$$y = \beta_0 + \beta_1 x_1 + \beta_2 x_1^2 + \beta_3 x_1^3 + \varepsilon$$

$$y = \beta_0 + \beta_1 \log x_1 + \beta_2 \log x_2 + \varepsilon$$

$$y = \beta_0 + \beta_1 x_1 + \beta_2 \frac{1}{x_1} + \varepsilon$$

$$y = \beta_0 + \beta_1 x_1 + \beta_2 x_2 + \beta_3 x_1 x_2 + \varepsilon$$

....

Estimates of parameters for such models are obtained by simple redefinitions of the independent variables. Thus, if it is desired to use the model

$$y = \beta_0 + \beta_1 x_1 + \beta_2 x_1^2 + \beta_3 x_1 x_2 + \beta_4 x_2^2 + \varepsilon$$

redefine new variables

$$z_1 = x_1$$
$$z_2 = x_1^2$$
$$z_3 = x_1 x_2$$
$$z_4 = x_2^2$$

and specify the model

$$y = \beta_0 + \beta_1 z_1 + \beta_2 z_2 + \beta_3 z_3 + \beta_4 z_4 + \varepsilon$$

for which the least-squares procedure can readily be used.

On the other hand, the following equations are examples of models that do not fall into the category of linear regression models:

$$y = \beta_0 + \beta_1 x_1 + \beta_1^2 x_1^2 + \beta_1 \beta_2 x_1 x_2 + \varepsilon$$
$$y = \beta_0 (\beta_1 x_1)(\beta_2 x_2)(\varepsilon)$$
$$y = \beta_0 + \beta_1 e^{\beta_2 x_2} + \varepsilon$$

It is possible to obtain least-squares estimates of coefficients of truly nonlinear models, but the normal equations for obtaining these estimates are not linear and therefore not easily solvable. For this reason, the general solution statement afforded by equation (2.22) and most of the statistical inference methodology discussed in Chapter 3 do not apply. Further discussion of these types of models is given in Chapter 6.

There are, however, some apparently nonlinear situations which can be linearized by simple redefinitions. For example, in the model

$$y = \beta_0 + \beta_1 x_1 + \frac{1}{\beta_2} x_2 + \varepsilon$$

one can redefine $\gamma_2 = 1/\beta_2$, and the resulting linear model is

$$y = \beta_0 + \beta_1 x_1 + \gamma_2 x_2 + \varepsilon$$

Another model may specify

$$y = \beta_0 + \beta_1 x_1 + \beta_2(x_2 + \beta_3 x_3) + \varepsilon$$

which is estimated by defining a model:

$$y = \gamma_0 + \gamma_1 x_1 + \gamma_2 x_2 + \gamma_3 x_3 + \varepsilon$$

The estimates of the parameters of the original model are

$$\hat{\beta}_1 = \hat{\gamma}_1$$

$$\hat{\beta}_2 = \hat{\gamma}_2$$

$$\hat{\beta}_3 = \frac{\hat{\gamma}_3}{\hat{\beta}_2}$$

2.10 PROBLEMS

1. Given the following set of artificial data:

	X_1	X_2	X_3	Y
	3	4	1	5
	4	7	8	7
	5	5	10	9
	5	5	0	11
	6	3	2	15
	7	6	9	13
Means	5	5	5	10

(a) Compute the *corrected* X'X and X'$\underline{y}$ matrices for the regression model $y = \beta_0 + \beta_1 x_1 + \beta_2 x_2 + \beta_3 x_3 + \varepsilon$.

(b) Obtain estimates of β_1, β_2 and β_3 by the forward Doolittle procedure.

(c) Obtain the inverse of X'X, and obtain $\underline{\hat{\beta}}$ by the formula

$$\underline{\hat{\beta}} = (X'X)^{-1} X'\underline{y}$$

(d) Obtain β_0 using formula (2.16).

(e) Obtain $\hat{\mu}_x$ for the first observation.

(f) Obtain *total* regression coefficients for the regression of y on x_1, x_2, and x_3, respectively.

(g) Obtain standardized (partial) regression coefficients, using equation (2.27). Compare magnitudes of actual and standardized coefficients.

2. Given the following artificial data:

	x_1	x_2	x_3	y
	4	5	7	5
	0	5	8	0
	4	5	5	9
	6	5	6	2
	4	0	1	8
	8	9	6	8
	4	7	6	4
	2	3	2	10
	4	6	13	8
Sums	36	45	54	54

(a) Estimate the coefficients of the multiple regression of the form

$$y = \beta_0 + \beta_1 x_1 + \beta_2 x_2 + \beta_3 x_3 + \varepsilon$$

using the *corrected* X'X matrix. Use forward Doolittle procedure.

(b) Estimate coefficients of the multiple regression using the model

$$y = \beta_1 x_1 + \beta_2 x_2 + \beta_3 x_3 + \varepsilon$$

and verify that $\Sigma(y - \hat{\mu}_{\underline{x}}) \neq 0$.

3. Given the following set of artificial data:

x_1	x_2	y
2	11	2
5	2	8
6	3	6
8	11	12
3	6	3
7	6	11
4	3	7

(a) Obtain estimates of regression coefficients using the *uncorrected* X'X and X'$\underline{y}$ matrix including the dummy variable.

(b) Obtain estimates of regression coefficients using the *corrected* SS and SP matrix, and then obtain β_0.

(c) Use the principle of residuals (Section 2.5) to obtain $\hat{\beta}_1$.

4. Data have been collected on the number of feeder cattle placed in feed lots for fattening (Southern Cooperative Series, 1965). We wish to estimate this number as a function of the price of range cattle (which constitute the *input* to the feeder operation), the price of slaughter cattle (which constitute the *output* of the feeder operation), and the price of corn (the main ingredient of the feed). The data are given in Table 2.6.

(a) Estimate the coefficients of a linear regression equation to predict placement of feeder cattle. Interpret the meaning of coefficients; are they reasonable?

(b) Using the equation of part (a), estimate cattle placement for the four quarters of 1960.

(c) Obtain standardized coefficients and comment on differences of magnitude of coefficients. The following is given for your benefit:

$$(X'X)^{-1} \text{ (corrected)} = \begin{bmatrix} 0.01257 & -0.01241 & 0.08856 \\ & 0.01991 & -0.02783 \\ & & 2.62870 \end{bmatrix}$$

TABLE 2.6 Data for Problem 4

Year	Quarter	Price of range cattle	Price of slaughter cattle	Price of corn	No. of head placed on feed (10^3)
1955	1	20.00	23.35	1.39	1744
	2	19.49	21.51	1.39	1586
	3	17.61	21.46	1.31	2099
	4	17.09	19.60	1.13	3791
1956	1	16.67	17.93	1.18	1851
	2	16.94	19.24	1.38	1538
	3	17.19	22.78	1.44	2387
	4	17.12	21.35	1.21	3862
1957	1	17.93	19.57	1.21	1723
	2	19.98	21.84	1.22	1583
	3	20.30	23.64	1.20	1903
	4	21.28	23.34	1.01	4062
1958	1	24.40	22.09	0.96	2182
	2	26.02	23.99	1.15	1765
	3	25.75	24.25	1.16	1965
	4	26.78	24.25	1.00	4449
1959	1	25.31	26.29	1.04	2295
	2	28.19	27.49	1.15	2015
	3	26.25	26.57	1.12	2374
	4	24.12	24.69	0.98	4369
1960	1	24.20	24.94	0.99	2416
	2	22.06	25.21	1.07	2013
	3	21.45	23.86	1.07	2360
	4	22.51	24.32	0.92	4738
1961	1	24.24	24.63	0.99	2419
	2	22.87	22.37	1.01	2028
	3	22.65	22.82	1.04	2765
	4	23.52	24.09	0.97	4870
1962	1	23.71	24.72	0.96	2505
	2	24.24	24.69	1.02	2187
	3	24.60	26.31	1.03	3036
	4	25.37	27.56	0.99	5351

$$X'\underline{y} = \begin{bmatrix} 23{,}053.4 \\ 22{,}752.0 \\ -2663.3 \end{bmatrix}$$

$\bar{x}_1 = 22.182$ $\bar{x}_2 = 23.461$ $\bar{x}_3 = 1.115$ $\bar{y} = 2694.7$

$s_1^2 = 11.234$ $s_2^2 = 5.4907$ $s_3^2 = 0.020948$ $s_y^2 = 1{,}215{,}850$

5. The following set of data illustrates a quadratic regression.

	x	y
	0	7.9
	1	12.0
	2	9.5
	3	11.3
	4	11.8
	5	11.3
	6	6.6
	7	4.2
	8	0.4
Sums	36	75.0

Obtain coefficient estimates for the model

$y = \beta_0 + \beta_1 x + \beta_2 x^2 + \varepsilon$

Calculate the $\hat{\mu}_x$-values and plot these and the original y-values against x.

6. Data have been collected on the distance covered by irrigation water (wetting front) in a furrow as related to time. An inspection of the data indicates that a straight-line regression will not adequately describe the relationship between distance y and

time x. A plot of the data suggests that a quadratic regression model

$$y = \beta_0 + \beta_1 x + \beta_2 x^2 + \varepsilon$$

will describe the relationship better. The data are

Obs.	Time, x (hr)	Distance, y (ft)
1	0.15	85
2	0.48	169
3	0.95	251
4	1.37	315
5	2.08	408
6	2.53	450
7	3.20	511
8	4.08	590
9	4.93	664
10	5.42	703
11	7.17	831
12	8.22	906
13	10.92	1075
14	11.92	1146
15	13.12	1222
16	15.78	1418
17	18.83	1641
18	21.22	1841
19	21.98	1864

Estimate coefficients of the quadratic regression model. To reduce computational effort the following quantities are provided (these are *not* corrected for the mean).

$\sum x = 154.35$ $\sum xy = 201454.6$

$\sum x^2 = 2183.7579$ $\sum x^2 y = 3314956$

$\sum x^3 = 37366.713$ $\sum y = 16063$

$\sum x^4 = 697035.84$ $\sum y^2 = 19053621$

7. Hamilton and Rubinoff (1963) discussed the prediction of species abundance (land plant species) in the Galapagos Archipelago. They list data for 17 islands of the Galapagos, including the following independent variables: x_1, area; x_2, elevation; x_3, distance to nearest island; x_4, distance from center of Archipelago; and x_5, area of the adjacent island. y is the observed species number. The data are given in Table 2.7. The estimates of the coefficients are

$\hat{\beta}_0 = 16.49$ $\hat{\beta}_1 = -0.027951$

$\hat{\beta}_2 = 0.07229$ $\hat{\beta}_3 = 3.092$

$\hat{\beta}_4 = 1.073$ $\hat{\beta}_5 = 0.09646$

Interpret the coefficients; are they reasonable? If not, suggest reasons for such results.

8. Given the following data:

x_1	x_2	y
2	3	1
4	5	7
5	6	10
7	5	8
7	5	9

verify, numerically, the equivalence of the (direct) multivariate and residual method of estimating the partial regression coefficient β_2.

TABLE 2.7 Prediction of Species Abundance in the Galapagos Archipelago

Island no. and name	x_1 Area (mi^2) AREA	x_2 Maximum elevation (ft) Height	x_3 Isolation (mi) nearest DSNEAR	x_4 Isolation (mi) center DSCENT	x_5 Area of adjacent island ARNEAR	No. of nearest island	y No. species observed
1. Culpepper	0.9	650	21.7	162	1.8	2	7
2. Wenman	1.8	830	21.7	139	0.9	1	14
3. Tower	4.4	210	31.1	58	45.0	5	22
4. Jervis	1.9	700	4.4	15	203.9	14	42
5. Bindlow	45.0	1125	14.3	54	20.0	12	47
6. Barrington	7.5	899	10.9	10	389.0	13	48
7. Gardiner	0.2	300	1.0	55	18.0	9	48
8. Seymour	1.0	500	0.5	1	389.0	13	52
9. Hood	18.0	650	30.1	55	0.2	7	79
10. Narborough	245.0	4902	3.0	59	2249.0	17	80
11. Duncan	7.1	1502	6.4	6	389.0	13	103
12. Abingdon	20.0	2500	14.3	75	45.0	5	119
13. Indefatigable	389.0	2835	0.5	0	1.0	8	193
14. James	203.0	2900	4.4	12	1.9	4	224
15. Chatham	195.0	2490	28.6	42	7.5	6	306
16. Charles	64.0	2100	31.1	31	389.0	13	319
17. Albemarle	2249.0	5600	3.1	17	245.0	10	325

Chapter 3

LINEAR MODELS: INFERENCE

3.1 INTRODUCTION

Chapter 2 discussed the implementation of the principle of least squares to estimate from sample data the parameters of a linear regression model of the form

$$y = \beta_0 + \beta_1 x_1 + \beta_2 x_2 + \cdots + \beta_m x_m + \varepsilon \qquad (3.1)$$

Although the resulting parameter estimates are useful, the use of such estimates without further analysis is not usually considered good statistics. It is usually of equal (if not greater) importance to make inferences regarding the adequacy of the model as a description of the population, as well as statements on the adequacy or reliability of the parameter estimates. Such aspects of statistical data analysis are usually performed by well-known principles of statistical inference and include the testing of hypotheses and making of confidence interval statements. Knowledge of these principles is presumed for readers of this chapter.

Some linear models contain many parameters. The existence of a multitude of parameters raises the problem of making separate or combined inferences regarding these parameters. There are two aspects of this problem:

1. The problem of proper significance levels in simultaneous inferences
2. The problem of inferences about individual parameters whose estimates are not independent of each other

The first aspect is similar to that of separation of means in the analysis of variance. Since this problem is not unique to regression analysis, it is not discussed further here. The second aspect is, however, a serious problem in many linear models analyses.

In the "usual" analysis of variance for balanced data, the estimates of the effects of the various experimental factors are indeed independent of each other, thus allowing for the easy computations and relatively trouble-free inferences on the respective treatment effects. However, if the data are not balanced, as in most regression analyses, the resulting parameter estimates are not independent and a different approach must be used.

The basic tool of statistical inference for linear model analysis is that of the partitioning of sums of squares, which is a natural consequence of the application of the least-squares estimating procedure. The various partitions of sums of squares are divided by appropriate degrees of freedom and used in an analysis of variance with the appropriate "F" tests. In the usual analysis of variance situations, this partitioning is accomplished by sums of squares formulas which are applied independently to each source of variation. However, when parameter estimates are not independent, it is not possible to directly partition the sum of squares into independent subdivisions corresponding to individual parameters or sets of parameters. Instead, a sum of squares associated with a particular parameter, or set of parameters, is obtained by computing residual sums of squares both with *and* without the parameter(s) in question and using the difference between these two as the appropriate sum of squares for the particular parameter(s). This procedure results in a sum of squares which provides a test involving only the parameter(s) in question.

A by-product of the nonindependence among parameter estimates is an often encountered contradiction among various inferences in a regression analysis. An extreme, but not uncommon, situation of this type consists of the rejection of the hypothesis that all β_i are zero (there is no regression) while not being able to reject the hypothesis of *any* partial β_i being equal to zero. This phenomenon is usually the result of a high degree of linear dependence or correlation (called *multicollinearity*) among independent variables. The existence of such dependence among independent variables implies that it is almost impossible to vary one of these variables while holding the others constant, hence the associated difficulty of isolating the *partial* regression coefficients. This problem is discussed in greater detail in Chapter 5.

3.2 SIMPLE LINEAR REGRESSION

The principles, computations, and resulting analysis of the partitioning of the sum of squares are first demonstrated here both by numerical example and by derivation of formulas for the one-variable linear regression model. Table 3.1 contains a sample data set of five pairs of observations of y and associated x-values; the data are plotted in Figure 3.1. Direct application of the estimation procedure from Chapter 2 produces estimates $\hat{\beta}_0 = 11.4$ and $\hat{\beta}_1 = -1.8$; hence, the estimated regression equation is

$$\hat{\mu}_x = 11.4 - 1.8x$$

providing the column of $\hat{\mu}_x$-values in Table 3.1 and the resulting regression line in Figure 3.1. We want to test $H_0\colon \beta_1 = 0$.

The partitioning of each observation according to the identity

$$(y - \bar{y}) = (y - \hat{\mu}_x) + (\hat{\mu}_x - \bar{y})$$

is demonstrated for the observation at x = 2 in Figure 3.1, and the corresponding numerical values for all observations are given in Table 3.1. The sums of squares of these values are given at the bottom of the table, and it is evident that

$$\sum(y - \bar{y})^2 = \sum(y - \hat{\mu}_x)^2 + \sum(\hat{\mu}_x - \bar{y})^2$$

The derivation of this equality is given later. The interpretation of the individual sums of squares is as follows:

TABLE 3.1 Partitioning of the Sums of Squares

Data		Estimates and partitions			
x	y	$\hat{\mu}_x$	$y - \bar{y}$	$\hat{\mu}_x - \bar{y}$	$y - \hat{\mu}_x$
1	10	9.6	4	3.6	0.4
2	7	7.8	1	1.8	-0.8
3	6	6.0	0	0.0	0.0
4	5	4.2	-1	-1.8	0.8
5	2	2.4	-4	-3.6	-0.4
Sums of squares			34	32.40	1.60

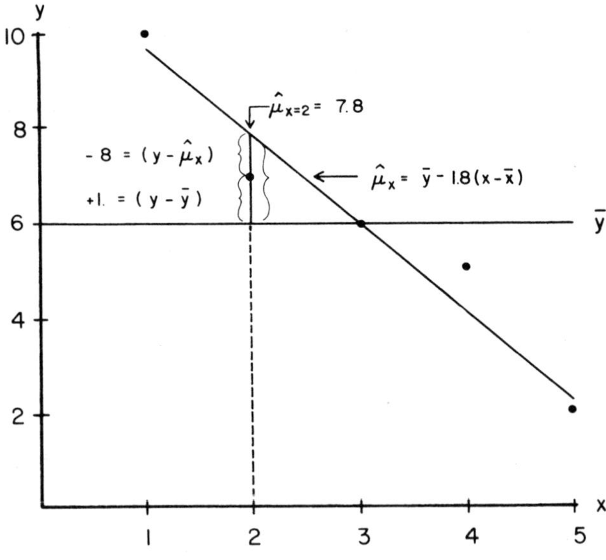

Figure 3.1 Partitions of Deviations.

1. $\Sigma(y - \bar{y})^2 = 34$ is the sum of squared deviations of the observed data from estimates obtained without assuming a (regression) model, since in the absence of a regression model, the model is $y = \mu + \varepsilon$, with μ estimated by $\bar{y}$. This quantity is the indicator of total variation; hence, it is called the *total* sum of squares (SST)
2. $\Sigma(y - \hat{\mu}_x)^2 = 1.60$ is the sum of squared deviations of observed data from estimates based on the (regression) model. This quantity is the indicator of the variation about the regression estimates; hence, it is called the *residual* (or *error*) sum of squares (SSE)
3. $\Sigma(\hat{\mu}_x - \bar{y})^2 = 32.40$ is the sum of squared differences between the (regression) model estimates and the "no model" estimates. This is called the sum of squares *due* to regression (SSR)

The algebraic derivation of the partitioning of sums of squares is straightforward and, in addition, produces some useful formulas. Starting with the previously stated identity

$$y - \bar{y} = (y - \hat{\mu}_x) + (\hat{\mu}_x - \bar{y})$$

square and sum to obtain

$$\begin{aligned}\sum(y - \bar{y})^2 &= \sum[(y - \hat{\mu}_x) + (\hat{\mu}_x - \bar{y})]^2 \\ &= \sum(y - \hat{\mu}_x)^2 + 2\sum(y - \hat{\mu}_x)(\hat{\mu}_x - \bar{y}) \\ &\quad + \sum(\hat{\mu}_x - \bar{y})^2 \qquad (3.2)\end{aligned}$$

In the middle term, substitute for $\hat{\mu}_x$ its estimate $\bar{y} + \hat{\beta}_1(x - \bar{x})$, and this term becomes[†]

[†]Note the use of the alternate statement of the model; see equation (2.5).

$$2\sum(y - \hat{\mu}_x)(\hat{\mu}_x - \bar{y}) = 2\sum[y - \bar{y} - \hat{\beta}_1(x - \bar{x})][\bar{y} + \hat{\beta}_1(x - \bar{x}) - \bar{y}]$$

since $\hat{\beta}$ is constant over the summation,

$$= 2\hat{\beta}_1 \sum[(y - \bar{y}) - \hat{\beta}_1(x - \bar{x})](x - \bar{x})$$

$$= 2\hat{\beta}_1[\sum(y - \bar{y})(x - \bar{x}) - \hat{\beta}_1 \sum(x - \bar{x})^2]$$

The least-squares estimate of $\hat{\beta}$ is

$$\hat{\beta}_1 = \frac{\sum(x - \bar{x})(y - \bar{y})}{\sum(x - \bar{x})^2}$$

hence,

$$\hat{\beta}_1 \sum(x - \bar{x})^2 = \sum(x - \bar{x})(y - \bar{y})$$

and the middle term of the original expression vanishes. Thus, it has been shown that

$$\sum(y - \bar{y})^2 = \sum(y - \hat{\mu}_x)^2 + \sum(\hat{\mu}_x - \bar{y})^2 \tag{3.2a}$$

or, in other words, that the total sum of squares is equal to the residual sum of squares plus the regression sum of squares. Note that the derivation of this identity required the use of the least-squares estimate of β_1; hence, the identity may not hold with other estimators.

The components of the sums of squares can be obtained without calculating individual values of $\hat{\mu}_x$. Substitute $\hat{\mu}_x = \bar{y} + \hat{\beta}_1(x - \bar{x})$ in the regression sum of squares, then

$$\begin{aligned} SSR &= \sum(\hat{\mu}_x - \bar{y})^2 \\ &= \sum[\bar{y} + \hat{\beta}_1(x - \bar{x}) - \bar{y}]^2 \\ &= \hat{\beta}_1^2 \sum(x - \bar{x})^2 \end{aligned} \tag{3.3a}$$

Alternatively, substitute the expression for $\hat{\beta}_1$, and

$$SSR = \frac{[\sum(x - \bar{x})(y - \bar{y})]^2}{\sum(x - \bar{x})^2} \tag{3.3b}$$

This is the most frequently used formula and is least subject to rounding error. Another equivalent formula is

$$SSR = \hat{\beta}_1 \sum(x - \bar{x})(y - \bar{y}) \tag{3.3c}$$

The residual sum of squares is most readily obtained by subtraction:

$$SSE = \sum(y - \hat{\mu}_x)^2 = SST - SSR \tag{3.4}$$

In order to ascertain that the above partitioning of sums of squares lends itself to the analysis of variance and the resulting F test for the hypothesis $\beta = 0$, it is instructive to develop the expressions that indicate what, in terms of the model parameters, is estimated by each of the sums of squares. It can be shown that the sample quantities

Total sum of squares (SST)

Regression sum of squares (SSR)

Residual sum of squares (SSE)

are estimates of

$$\begin{gathered}(n - 1)\sigma^2 + \beta_1^2 \sum(x - \bar{x})^2 \\ \sigma^2 + \beta_1^2 \sum(x - \bar{x})^2 \\ (n - 2)\sigma^2\end{gathered} \tag{3.5}$$

respectively, where σ^2 is the variance of the ε_i. These are known as *expected sums of squares*.

The coefficients of σ^2 in the expected sums of squares are the familiar degrees of freedom (d.f.), and dividing the sums of squares by degrees of freedom produces mean squares. The corresponding *expected mean squares* are given in Table 3.2. The residual mean square is the usual estimate of the variance of the ε_i and is often denoted $s^2_{y \cdot x}$ or simply s^2.

It can be seen that if $\beta_1 = 0$ (the usual null hypothesis), the ratio of the regression mean square to the residual mean square is a ratio of two (independent) estimates of σ^2, which is the basis for the F distribution with 1 and n - 2 degrees of freedom. If $\beta_1 \neq 0$, a positive quantity is added to the numerator of the ratio, hence the justification for the rejection region on the upper tail of the F distribution. The total mean square is not usually calculated. The format of the resulting analysis of variance table is presented in Table 3.3.

For the example of Table 3.1, SST = 34, SSR = $(-18)^2/10 = 32.4$, and SSE (by subtraction) = 34 - 32.4 = 1.6; these quantities obviously agree with the directly computed results (Table 3.1). Finally, MSR = 32.4 and MSE = 1.6/3 = 0.53; hence F = 60.75, leading to the rejection of the null hypothesis.

The null hypothesis H_0: $\beta_1 = \beta_{1*}$ may also be tested by the following t-statistic:

$$t_{(n-2)} = \frac{\hat{\beta}_1 - \beta_{1*}}{\sqrt{MSE/\sum(x - \bar{x})^2}}$$

TABLE 3.2 Expected Mean Squares

Source	d.f.	Expected mean squares
Total	n - 1	$\sigma^2 + \beta_1^2 \sum(x - \bar{x})^2/(n - 1)$
Regression	1	$\sigma^2 + \beta_1^2 \sum(x - \bar{x})^2$
Residual	n - 2	σ^2

TABLE 3.3 Analysis of Variance Format

Source	d.f.	SS	MS	F
Total	n - 1	$SST = \sum(y - \bar{y})^2$	--	
Regression	1	$SSR = \dfrac{[\sum(x - \bar{x})(y - \bar{y})]^2}{\sum(x - \bar{x})^2}$	MSR = SSR	MSR/MSE
Residual	n - 2	SSE = SST - SSR	$MSE = \dfrac{SSE}{(n - 2)}$	

It is easy to show that for the case $\beta_{1*} = 0$, this statistic is algebraically equivalent to the F-statistic in Table 3.3 [remember, $t^2_{(n-2)} = F(1, n - 2)$].

3.3 THE CORRELATION COEFFICIENT

A statistic associated with regression is the well-known correlation coefficient. The most common form of this statistic,

$$r = \frac{\sum(x - \bar{x})(y - \bar{y})}{\sqrt{\sum(x - \bar{x})^2 \sum(y - \bar{y})^2}} = \frac{\text{covariance }(x, y)}{s_x s_y} \qquad (3.6)$$

The quantity r lies between -1 and 1, and takes these extreme values only if x is an *exact* linear function of y.

The relationship between the correlation coefficient and regression is easily demonstrated. Squaring r and recombining terms,

$$r^2 = \frac{[\sum(x - \bar{x})(y - \bar{y})]^2}{\sum(x - \bar{x})^2} \frac{1}{\sum(y - \bar{y})^2}$$

$$= \frac{SSR}{SST}$$

It can further be shown that

$$r^2 = \frac{F}{n - 2 + F} \qquad (3.7)$$

or

$$F = \frac{(n - 2)r^2}{(1 - r^2)} \qquad (3.8)$$

where F is the F-value calculated from the analysis of variance (Table 3.3). Hence, the correlation coefficient is simply a function of the sample size and the test of the hypothesis $\beta = 0$ for linear regression.

It is readily seen that the correlation coefficient is symmetric with respect to x and y: it is the same if y is the *independent* variable and x is the *dependent* variable. Thus correlation can be used as a test statistic for a linear relationship between two variables regardless of which exhibits the random component.

The correlation coefficient can also be interpreted as the (absolute value) correlation between the observed and least-squares estimated y-values; that is

$$|r| = \frac{\sum(\hat{\mu}_x - \bar{y})(y - \bar{y})}{\sqrt{\sum(\hat{\mu}_x - \bar{y})^2 \sum(y - \bar{y})^2}} \tag{3.9}$$

3.4 MULTIPLE REGRESSION

It has already been pointed out that in multiple regression the existence of several parameters creates some difficulties for the inference process. The most commonly used procedure is first to examine the effectiveness of the entire model and then to examine the effectiveness of the several portions (separate parameters) of the model.

3.5 TEST FOR THE ENTIRE MODEL

The procedure for testing the existence of the simple linear regression extends readily to the multiple regression case. Specifically, it is desired to test the hypothesis

$$H_0: \underline{\beta} = 0$$

against

$$H_1: \underline{\beta} \neq 0$$

in the sense that any one or more elements of $\underline{\beta}$ are not equal to zero. As in the linear regression model, this is accomplished by the partitioning of the sum of squares:

$$SST = \sum(y - \bar{y})^2$$
$$SSR = \sum(\hat{\mu}_x - \bar{y})^2$$
$$SSE = \sum(y - \hat{\mu}_x)^2$$

where now

$$\hat{\mu}_x = \hat{\beta}_0 + \hat{\beta}_1 x_1 + \hat{\beta}_2 x_2 + \cdots + \hat{\beta}_m x_m$$

The fact that SST = SSR + SSE is demonstrated by a straightforward extension of simple linear regression. The computational formulas are more readily derived from the matrix form:

$$\underline{\hat{\mu}} = X\underline{\hat{\beta}}$$

where $\underline{y}$ and X are matrices of the observed variables, recorded as deviations from their respective sample means as defined in formula (2.24). The residuals from the estimated regression are

$$\underline{\hat{\varepsilon}} = \underline{y} - X\underline{\hat{\beta}}$$

and the sum of squared residuals is

$$\begin{aligned}\underline{\hat{\varepsilon}}'\underline{\hat{\varepsilon}} &= (\underline{y} - X\underline{\hat{\beta}})'(\underline{y} - X\underline{\hat{\beta}}) \\ &= \underline{y}'\underline{y} - \underline{\hat{\beta}}'X'\underline{y} - \underline{y}'X\underline{\hat{\beta}} + \underline{\hat{\beta}}'X'X\underline{\hat{\beta}}\end{aligned}$$

Insert the estimate $\underline{\hat{\beta}} = (X'X)^{-1}X'\underline{y}$ in the last term, which cancels with the second term; hence,

$$\underline{\hat{\varepsilon}}'\underline{\hat{\varepsilon}} = \underline{y}'\underline{y} - \underline{y}'X\underline{\hat{\beta}} \tag{3.10a}$$

or

$$\underline{\hat{\varepsilon}}'\underline{\hat{\varepsilon}} = \underline{y}'\underline{y} - \underline{\hat{\beta}}'X'\underline{y} \tag{3.10b}$$

From this it follows that since

$$\sum \hat{\varepsilon}^2 = \text{residual sum of squares}$$

and

$$\sum(y - \bar{y})^2 = \text{total sum of squares}$$

then

$$\sum_i^m \beta_i[\sum^n (x_i - \bar{x}_i)(y - \bar{y})] = \text{sum of squares due to regression} \qquad (3.10c)$$

Calculation of this quantity is straightforward; it is simply the sum of the products of the β_i and the already calculated $\Sigma(x_i - \bar{x}_i) \times (y - \bar{y})$. This quantity appears, at first glance, to be a generalization of the SSR for one-variable regression (3.3c), and in fact reducing (3.10c) to one independent variable reproduces the corresponding one-variable regression formula. It would then appear reasonable to assume that the individual terms in (3.10c) should provide sums of squares for testing individual β, but this is not the case, and it must be emphatically stated that *the individual terms have no meaning* (unless X'X is diagonal); although they may look like sums of squares, they can actually be negative.

As in the one-variable case, SSR can be calculated in several ways:

$$\text{SSR} = \underline{\hat{\beta}}'X'\underline{y} = \underline{\hat{\beta}}'X'X\underline{\hat{\beta}} = \underline{y}'X(X'X)^{-1}X'\underline{y} \qquad (3.11)$$

These latter expressions have more theoretical than practical significance, since they are more difficult to compute. Finally, it should be noted that if $\underline{y}$ and X are not corrected for the mean, and where X includes the dummy variable, then $\underline{\hat{\beta}}'X'\underline{y}$ will be the sum of squares due to regression *including* the intercept.

In Section 3.2 it was shown how the expected mean squares related the various sums of squares to the model parameters of a simple linear regression model. The equivalent expressions for multiple regression are given in Table 3.4. It is apparent that the ratio MSR/MSE gives the same type of result obtained in linear regression; in fact, the quantity $\underline{\beta}'(X'X)\underline{\beta}$ reduces to $\beta_1^2 \sum(x_1 - \bar{x}_1)^2$ if the dimension of the regression is reduced to unity. The quantity $\underline{\hat{\beta}}'(X'X)\underline{\hat{\beta}}$ is a positive definite quadratic form (see Section 1.7), which means that it equals zero only if $\underline{\beta} = 0$ and is positive for any $\underline{\beta}$ containing at least one nonzero value. Hence, the ratio MSR/MSE is the ratio of two independent estimates of σ^2 and follows the F distribution *only* if all elements of $\underline{\beta}$ are zero. Otherwise, the ratio will have a larger numerator and will follow a noncentral ("pulled to the right") F distribution, hence the usual justification for comparing this ratio to the F-distribution as the test of the hypothesis $\underline{\beta} = 0$. Note, however, that the rejection of the hypothesis gives no clue as to which coefficients contribute to that rejection; this is investigated by procedures to be discussed later.

We continue using the example of the cost of operating an auction market (Section 2.5, Table 2.3). The corrected sum of squares of the dependent variable (volume) is

$$SST = \sum(y - \bar{y})^2 = \sum y^2 - \frac{(\sum y)^2}{n} = 8467.80$$

According to (3.11), the sum of squares due to regression is

$$\begin{aligned} SSR &= \sum_i^m \hat{\beta}_i \sum^n (x_i - \bar{x}_i)(y - \bar{y}) \\ &= (3.21549)(1442.19) + (1.61316)(387.72) \\ &\quad + (0.81488)(994.15) + (0.80259)(2322.29) \\ &= 7936.75 \end{aligned}$$

The residual sum of squares is obtained by subtraction:

$$\begin{aligned} SSE &= SST - SSR \\ &= 8467.80 - 7936.75 \\ &= 531.05 \end{aligned}$$

This quantity can be verified directly as $\Sigma(y - \hat{\mu}_x)^2$, but this is computationally tedious. (See, however, problem 1 at the end of this chapter.)

TABLE 3.4 Analysis of Variance

Source	d.f.	SS	MS	E(MS)
Total	$n - 1$	SST	--	
Regression	m	SSR	MSR	$\sigma^2 + \underline{\beta}'(X'X)\underline{\beta}$
Residual	$n - m - 1$	SSE	MSE	σ^2

TABLE 3.5 Analysis of Variance

Source	d.f.	SS	MS	F
Total	18	8467.80		
Due to regression	4	7936.75	1984.19	52.31
Residual	14	531.05	37.93	

These quantities are used in the analysis of variance as given in Table 3.5. The resulting value of F (52.31 with 4 and 14 degrees of freedom) is obviously far into the 1% critical region; hence, the null hypothesis $\underline{\beta} = 0$ is readily rejected.

3.6 MULTIPLE CORRELATION

In Section 3.3, it was noted that the correlation coefficient provided a convenient index useful for testing the statistical significance of one-variable regression. Similarly, a multiple correlation can be defined analogously to expression (3.9):

$$R = \frac{\sum(\hat{\mu}_x - \bar{y})(y - \bar{y})}{\sqrt{\sum(\hat{\mu}_x - \bar{y})^2 \sum(y - \bar{y})^2}} \tag{3.12}$$

where $\hat{\mu}_x$ are the estimates resulting from the multiple regression model. This is known as the *multiple correlation* coefficient; the associated R^2, the *coefficient of determination*, is more frequently used; it is often stated as a percentage. A derivation very similar to that in Section 3.3 shows that, in this case also,

$$R^2 = \frac{SSR}{SST} \tag{3.13}$$

from which can be obtained the equivalent statistical significance test:

$$F = \frac{(n - m - 1)R^2}{m(1 - R^2)} \tag{3.14a}$$

or

$$R^2 = \frac{mF}{(mF + n - m - 1)} \tag{3.14b}$$

The coefficient of determination affords a convenient and simple description for the goodness of fit of a regression model. It is, in fact, so convenient that it is frequently misused; hence, it is advisable to discuss some cautions in the use of this statistic.

a. As formula (3.13) clearly shows, R^2 is a *relative* quantity, that is, it indicates how large SSR is relative to SST. In some situations, data may be quite variable and a large R^2 may not indicate a very good fit, while in more controlled situations, a relatively small R^2 may indicate a rather good fit. An example of the former situation is instructive.

Table 3.6 gives data on 16 utility companies relative to their capital, labor, and energy requirements for generating electrical power. A linear regression analysis produces an R^2 of 92.20%, certainly a respectable value. However, another description of the adequacy of the model is the residual standard deviation. For this example, this is 1.029×10^6 kWh, whose magnitude can be appreciated by looking at the predicted values (last column), which show that, for example, the fourth firm is estimated to have a power generation of 7.585×10^6 kWh and misses the true value by over 2×10^6 kWh, a difference of over 25%. Several other firms have estimates that are nearly as poor. Thus it can be seen that the R^2 is indeed a relative quantity; the large R^2 is largely due to the fact that large power companies, such as number 13, generate more power than smaller companies, such as number 9.

Another way to look at the inadequacy of R^2 is to note that using only the obviously important factor of energy consumption as an independent variable produces an R^2 value of 91.93%. If the primary purpose of the analysis is to study the role of capital and labor, it is seen that of the 8.07% $(1 - R^2)$ of variation not accounted for by energy consumption, only 3.35% (0.27/8.07) is accounted for by these other factors. In other words, beyond the already established fact that across a wide range of utility company sizes, electric power generation is closely related to energy input, capital and labor inputs contribute rather little for these data.

b. If the regression is required to pass through the origin (see Section 2.8), most computer printouts will use as the total sum of squares the uncorrected sum of squares for the dependent variable. Since this is usually a much larger number than the more frequently used corrected sum of squares, it can easily be seen that the corresponding R^2-value will tend to be quite large.

TABLE 3.6 Data on Utility Companies

Firm	Capital ($\$10^6$)	Labor (10^3)	Energy (10^9 BTU)	Output (10^6 kWh)	Predicted output (10^6 kWh)
1	98.288	0.386	13.219	1.270	1.826
2	255.068	1.179	49.145	4.597	4.961
3	208.904	0.532	18.005	1.985	2.421
4	528.864	1.836	75.639	9.897	7.585
5	307.419	1.136	52.234	5.907	5.384
6	138.283	1.085	9.027	1.832	1.116
7	418.883	2.390	41.676	4.865	3.926
8	247.439	1.356	31.244	2.728	3.214
9	19.478	0.115	1.739	0.125	0.770
10	537.540	2.591	104.584	9.685	9.777
11	605.507	2.789	82.296	8.727	7.790
12	174.765	0.933	21.990	2.239	2.465
13	946.766	4.004	125.351	10.077	11.733
14	296.490	1.513	43.232	4.477	4.317
15	645.690	2.540	75.581	7.037	7.420
16	288.975	1.416	42.037	3.507	4.250

c. If no regression exists in the population, the expected value of F is near unity and the corresponding value of R^2 is approximately m/n. Hence, a large value of R^2 means very little if m/n is large. For such situations it may be more appropriate to use the "adjusted" coefficient of determination:

$$\bar{R}^2 = 1 - \frac{n - 1}{n - m - 1}(1 - R^2) \tag{3.15}$$

This quantity represents the reduction in the *mean* square rather than the *sum* of squares, and in this manner it "adjusts" for the loss in degrees of freedom; for small values of m/n, $\bar{R}^2$ will tend to R^2. It is possible for $\bar{R}^2$ to be negative.

3.7 TEST FOR SUBSET OF THE MODEL

In the one-variable regression case, the test of significance (Table 3.3) is the only test needed (except in the rare event a test on β_0 is needed). In the multiple regression case, however, the analysis of variance test (Table 3.4) only indicates that one or more β is not zero; it does not indicate which these are. We require additional tests to answer this question.

Specifically, consider a subset of one or more of the individual regression coefficients; denote this subset, as a portion of $\underline{\beta}$, by $\underline{\beta}_1$ and the remaining subset of the coefficients by $\underline{\beta}_2$.[†] The matrix of corresponding x-values will be partitions of X, denoted X_1 and X_2. It is desired to develop a test for the hypothesis $\underline{\beta}_2 = 0$, where this hypothesis is rejected if one or more elements of $\underline{\beta}_2 \neq 0$.

In order to be meaningful, the test must be on the partial coefficients, that is, the contribution of $\underline{\beta}_2$ over and above that

---

[†]Since the numbering of variables is arbitrary, $\underline{\beta}_1$ and $\underline{\beta}_2$ may contain any desired variables.

provided by $\underline{\beta}_1$. Thus the testing procedure is based on the partitioning of sums of squares according to models with and without the regression parameters in question. In other words, consider partitions according to the model

$$\underline{y} = X_1\underline{\beta}_1 + X_2\underline{\beta}_2 + \underline{\varepsilon} \tag{3.16}$$

and the model

$$\underline{y} = X_1\underline{\beta}_1^* + \underline{\varepsilon}^* \tag{3.17}$$

(the asterisks denote that the parameters are not necessarily the same in the two models). The difference between the residual sum of squares of the two models is the basis of the significance test for H_0: $\underline{\beta}_2 = 0$. A commonly used phrase for this procedure is to define this as testing for the statistical significance of $\underline{\beta}_2$, *adjusted* for the effect of $\underline{\beta}_1$. Note that to consider the model

$$y = X_2\underline{\beta}_2^* + \varepsilon^{**}$$

and perform a test on $\hat{\underline{\beta}}_2^*$ is an entirely different matter, as it is equivalent to making tests of the individual total regression coefficients (see Section 2.5). The direct application of the differencing principle is computationally tedious, since it involves computing two regressions, the entire multiple regression and that involving $\underline{\beta}_1$. An application of partitioned matrices (Section 1.6) provides a simpler method. Assume all variables already corrected for the mean (or put the dummy variable x_0 into X_1); then the partitioned model is written

$$\underline{y} = [X_1X_2]\begin{bmatrix}\underline{\beta}_1\\ \underline{\beta}_2\end{bmatrix} + \varepsilon \tag{3.18}$$

and the estimated coefficients,

$$\begin{bmatrix} \hat{\underline{\beta}}_1 \\ \hline \hat{\underline{\beta}}_2 \end{bmatrix} = \left[\begin{array}{c|c} X_1' X_1 & X_1' X_2 \\ \hline X_2' X_1 & X_2' X_2 \end{array}\right]^{-1} \begin{bmatrix} X_1'\underline{y} \\ \hline X_2'\underline{y} \end{bmatrix}$$

$$= \left[\begin{array}{c|c} C_{11} & C_{12} \\ \hline C_{21} & C_{22} \end{array}\right] \begin{bmatrix} \underline{g}_1 \\ \underline{g}_2 \end{bmatrix}$$

$$= \begin{bmatrix} C_{11}\underline{g}_1 + C_{12}\underline{g}_2 \\ C_{21}\underline{g}_1 + C_{22}\underline{g}_2 \end{bmatrix} \tag{3.19}$$

The regression model using *only* X_1,

$$\underline{y} = X_1\underline{\beta}_1^* + \varepsilon^*$$

provides the estimate

$$\hat{\underline{\beta}}^* = (X_1'X_1)^{-1}X_1'\underline{y} = (X_1'X_1)^{-1}\underline{g}_1 \tag{3.20}$$

Note that $(X_1'X_1)^{-1} \neq C_{11}$ [equation (1.11)].

The sum of squares due to the regression involving *all* variables is written

$$\text{SSR(all)} = \hat{\underline{\beta}}_1'\hat{\underline{\beta}}_2' \begin{bmatrix} \underline{g}_1 \\ \underline{g}_2 \end{bmatrix} = [\hat{\underline{\beta}}_1'\underline{g}_1 + \hat{\underline{\beta}}_2'\underline{g}_2] \tag{3.21}$$

Note that $\hat{\underline{\beta}}_1'\underline{g}_1$ and $\hat{\underline{\beta}}_2'\underline{g}_2$ have no meaning individually [see discussion of the formula (3.10c)]. The sum of squares due to the regression of the variables in X_1 *only* is

$$\text{SSR(all)} = [\hat{\underline{\beta}}_1'\hat{\underline{\beta}}_2'] \begin{bmatrix} \underline{g}_1 \\ \underline{g}_2 \end{bmatrix} = [\hat{\underline{\beta}}_1\underline{g}_1 + \hat{\underline{\beta}}_2'\underline{g}_2] \tag{3.22}$$

The sum of squares due to the $\underline{\beta}_2$ coefficients adjusted for $\underline{\beta}_1$ is obtained by subtraction:

$$\text{SSR}(\underline{\beta}_2 \text{ adj}) = \text{SSR(all)} - \text{SSR}(\underline{\beta}_1 \text{ only})$$
$$= \hat{\underline{\beta}}_1'\underline{g}_1 + \hat{\underline{\beta}}_2'\underline{g}_2 - \hat{\underline{\beta}}_1^{*\prime}\underline{g}_1 \tag{3.23}$$

Using formula (1.12), $\underline{\beta}_1^*$ can be defined involving only quantities obtained for the entire model estimates:

$$\hat{\underline{\beta}}_1^* = \hat{\underline{\beta}}_1 - C_{12}C_{22}^{-1}\hat{\underline{\beta}}_2$$

Substituting this in (3.23) (note that $C_{12}' = C_{21}$), one obtains

$$\text{SSR}(\underline{\beta}_2 \text{ adj}) = \hat{\underline{\beta}}_1'\underline{g}_1 + \hat{\underline{\beta}}_2'\underline{g}_2 - \hat{\underline{\beta}}_1'\underline{g}_1 + \hat{\underline{\beta}}_2'\underline{C}_{22}^{-1}\underline{C}_{21}\underline{g}_1 \tag{3.23a}$$

The first and third terms cancel. From (3.19),

$$\hat{\underline{\beta}}_2 = C_{21}\underline{g}_1 + C_{22}\underline{g}_2$$

hence,

$$C_{21}\underline{g}_1 = \hat{\underline{\beta}}_2 - C_{22}\underline{g}_2$$

Substituting this in (3.23a), one obtains

$$\text{SSR}(\underline{\beta}_2 \text{ adj}) = \hat{\underline{\beta}}_2'\underline{g}_2 + \hat{\underline{\beta}}_2'C_{22}^{-1}\hat{\underline{\beta}}_2 - \hat{\underline{\beta}}_2'C_{22}^{-1}C_{22}\underline{g}_2$$

The first and third terms cancel, and finally

$$\text{SSR}(\underline{\beta}_2 \text{ adj}) = \hat{\underline{\beta}}_2'C_{22}^{-1}\hat{\underline{\beta}}_2 \tag{3.24}$$

Certain features of this formula are of interest:

1. It is a function involving the $\hat{\underline{\beta}}_2$ portion of $\hat{\underline{\beta}}$ and the inverse of a *portion* of the entire inverse. In other words, it uses exclusively information obtained in the process of estimating the full model
2. It is a positive definite quadratic form; hence, it can never be negative
3. If $\underline{\beta}_2 = \underline{\beta}$ and $\underline{\beta}_1$ is empty, providing a test for the entire $\underline{\beta}$-vector, (3.24) is identical to (3.11)
4. If $\underline{\beta}_2$ contains only one variable, say β_i, then (3.24) reduces to

$$\mathrm{SSR}(\beta_i \text{ adj}) = \frac{\hat{\beta}_i^2}{c_{ii}} \tag{3.25}$$

where c_{ii} is the i-th diagonal element of C (see Section (3.7)

5. It can be shown that (3.24) is an estimate of $k(\sigma^2 + \underline{\beta}_2' C_{22}^{-1} \underline{\beta}_2)$, where k is the order of $\underline{\beta}_2$. Dividing by k, the degrees of freedom associated with this sum of squares, provides a mean square for an F test for H_0: $\underline{\beta}_2' C_{22}^{-1} \underline{\beta}_2 = 0$, which is satisfied only if $\underline{\beta}_2 = 0$. It can be shown, however, that the expected value of $\hat{\underline{\beta}}_2^* \underline{g}_2$ is a function of both $\underline{\beta}_1$ and $\underline{\beta}_2$ and is therefore not useful as the basis for a hypothesis test on $\underline{\beta}_2$.
6. By symmetry one can compute

   $$\mathrm{SSR}(\hat{\underline{\beta}}_1 \text{ adj}) = \hat{\underline{\beta}}_1' C_{11}^{-1} \hat{\underline{\beta}}_1$$

   but, just as the individual terms in $\hat{\underline{\beta}}'\underline{g}$ have no meaning,

   $$\mathrm{SSR}(\underline{\beta}_1 \text{ adj}) + \mathrm{SSR}(\underline{\beta}_2 \text{ adj}) \neq \mathrm{SSR}(\text{all})$$
7. By the use of the procedure used to obtain (3.26), it is, by definition, true that

   $$\mathrm{SSR}(\underline{\beta}_1 \text{ only}) + \mathrm{SSR}(\underline{\beta}_2 \text{ adj}) = \mathrm{SSR}(\text{all})$$

   or

   $$\mathrm{SSR}(\underline{\beta}_2 \text{ only}) + \mathrm{SSR}(\underline{\beta}_1 \text{ adj}) = \mathrm{SSR}(\text{all})$$

   These relationships are sometimes useful for saving computational effort. For example, to test the hypothesis H_0: $(\beta_2, \beta_3, \ldots, \beta_{10}) = 0$ for a 10-variable equation, requires the inversion of a 9×9 matrix according to (3.24). Alternately, subtracting $\mathrm{SSR}(\beta_1 \text{ only})$ from SSR(all) gives the same results with less effort.
8. The procedure of computing partial regressions by residuals readily extends to multiple regression. Furthermore, the

statistical tests performed on the corresponding regression as if it were a total regression also provide the proper sums of squares for the test of the subset of partial regression coefficients.

Some of the partial (adjusted) sums of squares for individual coefficients and subsets can be obtained as a by-product of the forward portion of the abbreviated Doolittle matrix inversion procedure (Figure 1.1). The quantity $a'_{1g}b_{1g}$ is the sum of squares due to β_1 (only), the quantity $a'_{2g}b_{2g}$ is the sum of squares due to β_2 adjusted for β_1, the quantity $a'_{3g}b_{3g}$ is the sum of squares due to β_3 adjusted for β_1 and β_2, etc. From this it follows (by item 7), for example, that $\sum_{i=1}^{k} a'_{ig}b_{ig}$ is the sum of squares due to the first k coefficients (only) and $\sum_{i=\ell}^{m} a'_{ig}b_{ig}$ is the adjusted sum of squares due to the last $m - \ell + 1$ coefficients; $a'_{im}b_{im}$ is the fully adjusted sum of squares due to β_m. Note, however, that the variables must be inserted into the X'X matrix in a specific sequence in order to obtain specific adjusted subset sums of squares.

In the example of estimating the cost of operating auction markets (Section 2.5), the partial coefficients have been estimated, and the entire model was found to be statistically significant (Table 3.5). Cattle and calves are physically the larger units and are therefore costlier to handle, as evidenced by the larger coefficients associated with their numbers (Table 2.5). It is thus reasonable to ask whether the number of cattle and calves alone may be used to estimate operating cost or, conversely, whether the number of hogs and sheep contribute to cost. This is tested by

$$H_0: \begin{bmatrix} \beta_3 \\ \beta_4 \end{bmatrix} = \begin{bmatrix} 0 \\ 0 \end{bmatrix}$$

Using the data from Table 2.5 in formula (3.24) gives

$$SSR(\beta_3, \beta_4, \text{adj}) = [0.81488 \quad 0.80259]\begin{bmatrix} 0.00584200 & -0.00088954 \\ -0.00088954 & 0.00094989 \end{bmatrix}^{-1}$$

$$\times \begin{bmatrix} 0.81488 \\ 0.80259 \end{bmatrix}$$

$$= 1168.02$$

This sum of squares has 2 degrees of freedom, and, using the residual mean square as error, the test statistic is

$$F(2, 14) = \frac{1168.02/2}{37.93}$$

$$= 15.397$$

which leads to the rejection of that hypothesis at the 0.01 level. Thus sales of hogs and sheep do contribute to costs over and above the sales of cattle and calves.

3.8 A SUBSET OF SIZE ONE: A SPECIAL CASE

The sum of squares due to a single partial coefficient, say β_i, has previously been defined as

$$SSR(\beta_i \text{ adj}) = \frac{\hat{\beta}_i^2}{c_{ii}} \tag{3.26}$$

and the resulting F test can be written

$$F(1, n - m - 1) = \frac{\hat{\beta}_i^2/c_{ii}}{s^2}$$

where s^2 is the *residual mean square*. It is a well-known fact that $F(1, \nu) = t^2(\nu)$; hence, the test statistic can be written

$$t_{(n-m-1)} = \pm \frac{\hat{\beta}_i}{\sqrt{c_{ii}s^2}} \tag{3.27}$$

Since the t test for the hypothesis H_0: $\theta = 0$, where the estimate $\hat{\theta}$ of the parameter θ is normally distributed, is of the form

$$t = \frac{\hat{\theta}}{\text{estimated standard error of } \hat{\theta}} \tag{3.27a}$$

it can be inferred that the estimated standard error of $\hat{\beta}_i$ is $\sqrt{c_{ii}s^2}$.

This relationship can also be derived by other methods which would, however, involve concepts beyond the scope of this book. The results of such a derivation are stated as follows:

> Under the usual assumptions, the estimate $\hat{\beta}_i$ of β_i is a normally distributed random variable with mean β_i and variance $c_{ii}\sigma^2$. Further, the covariance between estimates of two coefficients, $\hat{\beta}_i$ and $\hat{\beta}_j$, is $c_{ij}\sigma^2$. Thus the test for the hypothesis $\beta_i = 0$ is of the form
>
> $$z = \frac{\hat{\beta}_i}{\sqrt{c_{ii}\sigma^2}}$$
>
> where z is a sample from the *standard normal*. Estimating σ^2 by the residual mean square s^2 provides the previously stated results.

Using the results in Tables 2.5 and 3.5, the t-statistics for testing the individual coefficients are

$$H_0\colon \beta_1 = 0 \qquad t = \frac{3.21549}{\sqrt{(0.0046983)(37.93)}} = 7.617$$

$$H_0\colon \beta_2 = 0 \qquad t = \frac{1.61316}{\sqrt{(0.0191227)(37.93)}} = 1.895$$

$$H_0\colon \beta_3 = 0 \qquad t = \frac{0.81488}{\sqrt{(0.0058425)(37.93)}} = 1.731$$

$$H_0\colon \beta_4 = 0 \qquad t = \frac{0.80259}{\sqrt{(0.00094989)(37.93)}} = 4.228$$

To reject at 5% significance (two-tailed), a t-value $\geq |2.145|$ is required; hence, we cannot reject the nonexistence of β_2 or β_3, i.e., we cannot conclude that the number of calves or hogs, respectively, contributes to the cost of operating an auction market. This does not, however, necessarily imply that *both* of these should be dropped from the model. This appears to contradict the example test of the previous section; this type of phenomenon is discussed further in Chapter 5.

Note that F-statistics could have been used with identical results. The t-statistic does, however, have some specific advantages:

1. The t-statistic may be used for a one-tailed test, for example,

   H_0: $\beta_i > 0$ versus H_1: $\beta_i = 0$

2. The t-statistic may be used as the test of a specific value of β, for example,

   H_0: $\beta_i = \beta_{i0}$ versus H_1: $\beta_i \neq \beta_{i0}$

For example, assume that it is desired to calibrate a measuring device for a physical phenomenon (e.g., temperature). If the new device is correctly calibrated, then the regression model

$$y = \beta_0 + \beta_1 x + \varepsilon$$

where

y = measurement observed on the new device

x = measurement observed on the old (standard) device

should have

$$\beta_0 = 0 \quad \text{and} \quad \beta_1 = 1$$

hence, the most logical test for validating the new device is H_0: $\beta_1 = 1.0$.

3. The above can be used to test a hypothesis regarding the value of a linear function of the β. A specific application of this is given in Section 3.10.

3.9 PARTIAL CORRELATION

In previous sections, a correlation coefficient was found to provide a useful indicator of the degree of a linear relationship. The corresponding statistic for a partial regression coefficient is the partial correlation coefficient, which can be defined in terms of residuals as follows.

Let $\hat{\underline{\delta}}^{(1)}$ be the vector of the estimated residuals for the linear regression based on the model $\underline{x}_1 = X\underline{\beta}^{(1)} + \underline{\delta}^{(1)}$, where X is a matrix of observations on a set of variables arbitrarily labeled $x_3, x_4, \ldots, x_p$. Similarly, let $\hat{\underline{\delta}}^{(2)}$ be the vector of estimated residuals from the regression $\underline{x}_2 = X\underline{\beta}^{(2)} + \underline{\delta}^{(2)}$. Then the simple correlation between the corresponding elements of $\hat{\underline{\delta}}^{(1)}$ and $\hat{\underline{\delta}}^{(2)}$, denoted $r_{12 \cdot 34 \cdots p}$, is the partial correlation of x_1 and x_2 independent of (adjusted for) $x_3, x_4, \ldots, x_p$.

Relabeling x_1 as a dependent variable allows the partial correlation to become the correlation measuring the degree of linear relationship corresponding to the partial regression of x_1 on x_2, adjusted for all other variables. As in the case of all correlations, the partial correlation is symmetric; hence, the above correlation also corresponds to the partial regression of x_2 on x_1. The square of the partial correlation is also a ratio of sums of squares: define

SSR = sum of squares due to β_i adjusted for all other variables

SST = total sum of squares of y, *adjusted* for all *other* variables, i.e., the *residual* after fitting all other variables.

Then SSR/SST is the square of the partial correlation of y and x_i, independent of all other variables.

The following comments on computations and usages of the partial correlation coefficient are of interest.

1. A partial correlation independent of a single variable is called a *first-order partial correlation*. The partial correlation between, say, x_i and x_j independent of x_k is

denoted by $r_{ij \cdot k}$ and can be obtained from simple correlations by the formula

$$r_{ij \cdot k} = \frac{r_{ij} - r_{ik}r_{jk}}{\sqrt{(1 - r_{ik}^2)(1 - r_{jk}^2)}} \tag{3.28}$$

where r_{ij} is the simple correlation between x_i and x_j. This formula is derived by computing the correlation between residuals from simple linear regressions of x_i and x_j on x_k, respectively.

An equivalent relationship holds between first- and second-order partial correlations:

$$r_{ij \cdot k\ell} = \frac{r_{ij \cdot k} - r_{i\ell \cdot k}r_{j\ell \cdot k}}{\sqrt{(1 - r_{i\ell \cdot k}^2)(1 - r_{j\ell \cdot k}^2)}} \tag{3.29}$$

In this manner it is possible to "partial out" variables, one at a time, until one obtains $r_{ij \cdot (\text{all other variables})}$.

2. It is possible to obtain all (m - 2)-order partials (as above) more directly without going through m - 2 repeated applications of the above formula. Using the inverse C of the X'X matrix of all variables, the desired partial correlations are

$$r_{ij \cdot (\text{all other variables})} = \frac{-c_{ij}}{\sqrt{c_{ii}c_{jj}}} \tag{3.30}$$

This method of obtaining partial correlations from the C matrix is the same method used to obtain simple correlations from the elements of X'X matrix, except that the sign is reversed. Note that in order to calculate the correlation $r_{ij \cdot (\text{all other x variables})}$ by this procedure, it is necessary to compute the inverse of a matrix of sums of squares and cross products which includes the dependent variable.

3. The partial correlation coefficient can be used to test the significance of the corresponding partial regression

coefficient. The t test for H_0: $\beta_{ij} = 0$, where β_{ij} is the partial regression of x_i on x_j, is

$$t_{(n-m-1)} = \sqrt{(n - m - 1)\,\frac{r_{ij}^2}{1 - r_{ij}^2}} \qquad (3.31)$$

where r_{ij} is the partial correlation of x_i and x_j, independent of all other variables. Note that this also tests H_0: $\beta_{ji} = 0$.

4. Partial correlation coefficients can be used to obtain any corresponding partial regression coefficient:

$$\hat{\beta}_{ij} = r_{ij\cdot(\text{all other variables})}\sqrt{\frac{c_{jj}}{c_{ii}}} ,$$

where $\hat{\beta}_{ij}$ is the partial regression coefficient of x_i on x_j, and c_{ii} and c_{jj} are the elements of the inverse of X'X.

A related quantity, sometimes called a *part correlation*, is the ratio of the partial sum of squares (as above) to the *total* sum of squares. This quantity indicates the portion of the total variation which is accounted for by the corresponding partial coefficient. It is not frequently used and must not be confused with partial correlation.

3.10 CONFIDENCE INTERVALS

Another important aspect of statistical inference is that of making estimates. The point estimates of parameters were discussed in Chapter 2. This section is devoted to a discussion of procedures for making *confidence interval* statements.

The most commonly used confidence interval is based on the t-distribution [see formula (3.27a)] and has the form

$$\Pr\{(\hat{\theta} - t_{(1-\alpha/2)}s_{\hat{\theta}}) < \theta \leq (\hat{\theta} + t_{(1-\alpha/2)}s_{\hat{\theta}})\} = 1 - \alpha \qquad (3.32)$$

where

θ = parameter for which the confidence interval is to be constructed,

$\hat{\theta}$ = point estimate of θ,

$s_{\hat{\theta}}$ = estimated standard error of $\hat{\theta}$,

$t_{(1-\alpha/2)}$ = $(1 - \alpha/2)$ percentage point of the t-distribution

$1 - \alpha$ = probability that the statement is true

This formulation of the confidence interval estimate holds for any parameter whose estimate is normally distributed and for which the estimated variance is distributed as a function of χ^2.

Under the usual assumptions (Section 2.2), least-squares estimates of linear model parameters have these properties. Thus, using the results of Section 3.7, it is possible to construct a confidence interval for a regression coefficient:

$$\Pr\{(\hat{\beta}_i - t_{(1-\alpha/2)}\sqrt{c_{ii}s^2}) < \beta_i \leq (\hat{\beta}_i + t_{(1-\alpha/2)}\sqrt{c_{ii}s^2})\} = 1 - \alpha \tag{3.33}$$

A generalization of the above concerns confidence intervals for a linear function of several β such as, for example, $\beta_i - \beta_j$. Construction of such intervals requires the formula for the variance of a linear function of random variables. Define the linear function

$$z = \sum a_i y_i = \underline{a}'\underline{y}$$

where a_i are known constants and y_i are random variables; variances = $\sigma_i^2 = \sigma_{ii}$ and covariances = σ_{ij}, $i \neq j$. Then the variance of z is

$$V(z) = \sum\sum a_i a_j \sigma_{ij} = \sum_i a_i^2 \sigma_{ii} + 2 \sum_{i<j} a_i a_j \sigma_{ij} \tag{3.34a}$$

$$= \underline{a}' \Sigma \underline{a} \tag{3.34b}$$

where Σ is the matrix of variances and covariances of the y variables. Note that this is a quadratic form and cannot be negative, since Σ must be positive definite.

The linear function $(\beta_i - \beta_j)$ can be written $\underline{a}'\underline{\beta}$, where the vector $\underline{a}$ contains +1 in the i-th position, -1 in the j-th position, and zeros elsewhere. A few substitutions show that the estimated variance is

$$\hat{V}(\hat{\beta}_i - \hat{\beta}_j) = s^2(c_{ii} + c_{jj} - 2c_{ij})$$

A more useful application concerns confidence intervals for regression estimates. The estimate of the regression plane or line $(\hat{\mu}_x)$ can be written [see equation (2.5)]

$$\hat{\mu}_x = \bar{y} + \hat{\beta}_1(x_1 - \bar{x}_1) + \hat{\beta}_2(x_2 - \bar{x}_2) + \cdots + \hat{\beta}_m(x_m - \bar{x}_m)$$

For a specific set of x-values, say x_{i0}, the quantities $(x_{i0} - \bar{x}_m)$ are constants, while $\bar{y}$ and the $\hat{\beta}_i$ are random variables. Further, it can be shown that

$$\text{Variance of } \bar{y} = \frac{s^2}{n}$$

and

$$\text{Covariance of } (\hat{\beta}_i, \bar{y}) = 0 \qquad (i = 1, 2, \ldots, m)$$

Utilizing the formula for the variance of a linear function gives

$$\text{Var}(\hat{\mu}_{\underline{x}_0}) = s^2[\frac{1}{n} + \sum_i \sum_j (x_{i0} - \bar{x}_i)(x_{j0} - \bar{x}_j)c_{ij}] \tag{3.35a}$$

or in matrix notation

$$s^2\left[\frac{1}{n} + (\underline{x}_0 - \bar{\underline{x}})'C(\underline{x}_0 - \bar{\underline{x}})\right] \tag{3.35b}$$

where $\underline{x}_0 - \bar{\underline{x}}$ is a vector of the $(x_{i0} - \bar{x}_i)$-values and the x_{i0} refer to a specified set of x-values. The corresponding confidence interval statement is

$$\Pr\{\hat{\mu}_{\underline{x}_0} - t\sqrt{s^2[\frac{1}{n} + \sum\sum(x_{i0} - \bar{x}_i)(x_{j0} - \bar{x}_j)(c_{ij})]} \le \mu_{x_0}$$
$$< \hat{\mu}_{\underline{x}_0} + t\sqrt{s^2[\frac{1}{n} + \sum\sum(x_{i0} - \bar{x}_i)(x_{j0} - \bar{x}_j)(c_{ij})]}$$
$$= 1 - \alpha \qquad (3.36)$$

Note that letting all $x_{i0} = 0$ provides a confidence interval for β_0.

The interval for an estimate of a single observation for a given set of x-values, denoted by $\hat{y}_{\underline{x}_0}$, is obtained in the same manner. In this case,

$$\hat{y}_{\underline{x}_0} = \bar{y} + \hat{\beta}_1(x_{10} - \bar{x}_1) + \hat{\beta}_2(x_{20} - \bar{x}_2) + \cdots + \hat{\beta}_m(x_{m0} - \bar{x}_m) + \varepsilon$$

where ε is the residual which is known to have a variance of σ^2, which is estimated by s^2. Further, the ε are independent of $\bar{y}$ and also the β_i; thus, the estimated

$$\text{Var}(\hat{y}_{\underline{x}_0}) = s^2[1 + \frac{1}{n} + \sum_i \sum_j (x_{i0} - \bar{x}_i)(x_{j0} - \bar{x}_j)(c_{ij})] \qquad (3.37)$$

The corresponding confidence interval is thus like (3.36) with the addition of the 1 inside the parentheses.

Since C is positive definite, each of the confidence intervals is narrowest at $\underline{x}_0 = \bar{\underline{x}}$, at which point $\hat{\mu}_x = \bar{y}$. Intervals may be computed and used for any values of $\underline{x}_0$ within the range of observed data. Extrapolation outside this range is not recommended, as it makes the unwarranted assumption that the estimated regression relationship holds outside the region of observed data. Of course, we are only assuming that it holds inside the region, but we have some means of partial verification there.

It is important to note the different implications of the two confidence interval statements. The interval given by (3.36) implies with 1 - α confidence that the population regression surface (conditional population mean) lies within the calculated limits. The interval based on (3.37) implies with $1 - \alpha$ confidence that a single observation of y will lie within the calculated limits.

Substituting the x-values for the first market (Table 2.3) in equation (3.35a) gives a variance estimate of 9.0866, and using $t_{0.975}(14) = 2.145$ gives a 95% confidence interval from 27.425 to 40.357. In other words, with 95% confidence, we state that the true mean operating cost of markets with sales figures of the first market lies between 27.425 and 40.357. The corresponding 95% confidence interval for a single observation is from 19.185 to 48.599. The latter interval is, as expected, considerably wider. Note that, as usual, the probability applies to the limits, which are random variables.

3.11 PROBLEMS

1. Use data from problem 1 of Chapter 2 and do the following:
   (a) Compute for $\hat{\mu}_x$ each of the six observations and then obtain directly $\Sigma(\hat{\mu}_x - \bar{y})^2$.
   (b) Obtain SSR and SST by formula (3.10c) and do the analysis of variance to test the hypothesis $\underline{\beta} = 0$.
   (c) Test hypotheses (separately) that each of the *total* regression coefficients is zero.
   (d) Test (separately) that each of the *partial* regression coefficients is zero. Compare results with part (c).
   (e) Compute the correlation between y and $\hat{\mu}_x$ [as obtained in part (a)] and check the value with $\sqrt{SSR/SST}$.
   (f) Using only x_1 and x_2, do the multiple regression and compare with results of separate total regressions obtained in part (c). What is the difference between this and part (d)?
   (g) Use the forward portion of the abbreviated Doolittle and verify the equivalences of the $a'_{ig}b_{ig}$ to the various sums of squares (end of Section 3.6).

2. (a) Use data from problem 3 of Chapter 2 and complete all relevant hypothesis tests.

   (b) Invert the entire 3 × 3 (i.e., including y) matrix of corrected SS and SP and obtain the partial correlations $r_{y1 \cdot 2}$ and $r_{y2 \cdot 1}$.

3. Given the following data:

X_1	X_2	Y
1	4	2
2	6	3
3	6	4
2	4	5
3	4	6
4	6	7
3	3	8
4	4	9
5	8	10

and the following derived quantities:

$$X'X \text{ (corrected for mean)} = \begin{bmatrix} 12 & 8 \\ & 20 \end{bmatrix}$$

$$X'\underline{y} = \begin{bmatrix} 24 \\ 6 \end{bmatrix} \qquad \underline{y}'\underline{y} = [60]$$

$$\bar{x}_1 = 3 \qquad \bar{x}_2 = 5 \qquad \bar{y} = 6$$

(a) Estimate *total* regression equations of y on x_1 and x_2, respectively.

(b) Estimate the multiple regression equation of y on x_1 and x_2.

(c) Estimate the partial regression coefficient β_2 by regressing residuals of regressions of x_2 on x_1.

(d) Show graphically the relationship between the total and partial coefficient for x_2.

TABLE 3.7 Data for Problem 4

Individual	Year 1	Year 3	Year 5
1	1.0395	0.8513	0.8142
2	0.8949	0.8663	0.8176
3	0.9773	0.7959	0.7308
4	0.8646	0.9009	0.8129
5	0.8357	0.8028	0.7126
6	1.2212	1.0000	1.0000
7	0.9763	0.9969	0.8768
8	1.0000	0.8470	0.8927
9	1.0000	0.8921	0.8102
10	0.8096	0.8802	0.7513

4. Ten individuals were given a test for a learned manual skill, and then the test was repeated 3 and 5 years later. The scores of the test are given in Table 3.7. A measure of learning and retention would be the ability to predict year 5 performance from performance in years 1 and 3. Test the hypothesis that nothing was learned or retained.

5. It is desired to predict scholastic success of a certain group of incoming freshmen at college. The following information is available for a sample of 88 such freshmen:

   1. Aptitude test score: verbal
   2. Aptitude test score: arithmetic
   3. Quartile of rank in high school (1 is highest, etc.)
   4. Grade point ratio (GPR), 4-point scale, first year in college (dependent variable)

   The sums and corrected sums of squares and cross products are given in Table 3.8.

TABLE 3.8 Data for Problem 5

	Verbal	Arithmetic	Rank	GPR
		Sums		
	37435	42370	188	188.28
		SS and SP		
Verbal	639527	470148	-2284.77	2173.57
Arithmetic		914768	-4281.82	3511.20
Rank			72.3636	-35.2535
GPR				47.4443
		Inverse[a]		
Verbal	2.51410	-1.27321	4.04190	--
Arithmetic		2.15671	87.4149	--
Rank			19119.1	--

[a]Elements multiplied by 10^6.

(a) Provide the least-squares estimating equation for predicting GPR from the test scores and rank. Test all relevant hypotheses.

(b) Psychologist Jones maintains that test scores do *not* help in predicting scholastic success. Do the appropriate test to either support or reject his theory.

(c) As registrar you decide to admit students in your college only if their predicted GPR is greater than 2.0. Give your decision on these students whose scores are given in Table 3.9.

(d) As president of the college you want to admit students for whom the 90% confidence interval for the individuals' scores for GPR includes 2.0. Compute this interval for each of these students. Discuss the implication of this policy versus that of the registrar. Note: Since the computations for this are tedious, the value of the quadratic form in equation (3.35b) is given in the column labeled Q.F. in Table 3.9.

TABLE 3.9 Data for Problem 5

Student	Verbal	Arithmetic	Rank	Q.F.
15	490	609	2	0.02039
16	468	635	1	0.03112
47	280	301	3	0.04366
78	431	662	1	0.05428

Chapter 4

THE "RANDOM" ERROR

Chapters 2 and 3 discussed the "classical" least-squares procedures for analyzing data according to the linear model

$$\underline{y} = X\underline{\beta} + \underline{\varepsilon}$$

subject to the usual assumptions:

1. The elements of $\underline{\beta}$ are constants (parameters)
2. The elements of X are predetermined and measured without error
3. The elements of $\underline{\varepsilon}$ are independently distributed random variables with the mean 0 and variance σ^2
4. The distribution of the elements of $\underline{\varepsilon}$ must be normal if t and/or F tests are to be used.

Assumption 1 is normally fulfilled for most practical problems associated with regression analysis. Violations of this assumption may be associated with situations described as nonstationary stochastic processes or other dynamic processes or models. Methods for analyzing such situations are beyond the scope of this book [see, e.g., Bhat (1972) and Box and Jenkins (1976)].

Assumption 2 would appear to be fulfilled primarily in the case of controlled experiments where x-values describe chosen levels or conditions of experimental factors. Data from such experiments are often analyzed by analysis of variance techniques. However, regression techniques are frequently used when the data simply "occur," i.e., when controls necessary to conduct a designed experiment cannot be exercised. In such cases, assumption 2 would seem to be in doubt.

For practical purposes, however, this does not prohibit the use of regression techniques if

a. It is not desired to make inferences on the independent variables
b. The dependent variable has no influence on the values of any of the independent variables
c. The independent variables are measured without error

Conditions a and b are usually met in practical situations; condition c is often not met but creates major problems only if the measurement error is large relative to the total variation of the variables. Since this is not often the case and since the treatment of measurement errors is involved and difficult, it will not be discussed further here [refer, e.g., to Johnston, (1972, Chapter 6)].

Assumptions 3 and 4 are, however, frequently violated for many practical problems. Fortunately, it has been shown that regression analysis procedures are relatively "robust," in that moderate departures from these assumptions do not drastically affect the validity of results obtained. However, knowledge of the effects of nonfulfillment of these assumptions is useful, since occasionally the effects are serious and, for some situations, there exist techniques which can, at least partially, overcome some of these deficiencies.

In this chapter are discussed three aspects of assumption 3:

Specification error

Autocorrelation

Heteroskedasticity

4.1 SPECIFICATION ERROR

The least-squares methodology discussed in the previous chapters uses the residuals from estimated values $(y - \hat{\mu}_x)$ to estimate the variance of the random error. If the model correctly describes the data, these residuals provide the proper estimate. However, in many practical applications, the correct model is not precisely known and may therefore be incorrectly specified; this is called *specification error*. For example, some independent variables may not have been measured because they were thought not to be relevant. It is also possible to have used incorrect functional forms of some independent variables; possibly the square of a variable should be included. Alternately, too many independent, hence irrelevant, variables may have been included in an attempt to avoid specification error.

It can be shown that the omission of terms in the regression model causes the residual mean square to be inflated and the coefficient estimates to be biased (Rao, 1971). Using too many variables in the model does not cause these effects, but creates other problems which are discussed in Chapter 5.

There is no universally applicable method for detecting specification errors. Nevertheless, critical examinations of results of regression analyses and proper use of other information may be used as a guide for detecting specification errors.

Statistically sound methods for ascertaining the existence of specification errors are available if there exist estimates (or other knowledge) of the "true" error variance. The most frequently available source for such statistics is replicated measurements, that is, repeated observations for identical values of all independent variables. This is particularly applicable for the detection of, say, the need for nonlinear terms. In other situations, results from previous experiments or data collections can provide such knowledge.

A more subjective method which requires no prior information consists of plotting the residuals from the regression estimates, i.e., the values of $y - \hat{\mu}_x$. In most statistical computer packages, such residuals can be plotted against any variable, dependent or independent, as well as the estimated values themselves. A correct model specification should result in a random scatter of these points, regardless of how they are plotted. A nonrandom pattern may indicate deviation from correct specification. The definition of *nonrandom* pattern is, of course, somewhat obscure, and apparently *random* patterns may exist in spite of specification errors. Consequently, examination of residuals is not *guaranteed* to detect such errors.

Residuals plots are also useful for the detection of unusual or suspect observations, that is, individual residuals which appear to lie outside the range of most other observations; these are called *outliers*. Formal statistical tests for outliers which may lead to the discarding of observations are available [e.g., Anscombe (1960)] and will not be discussed here. However, such formal tests are not required to simply investigate suspect observations which may result from possible model deficiencies. In such cases, it may be useful to identify observations corresponding to suspect residuals and see whether a pattern among the independent variables may suggest misspecification. For example, large and opposite-signed residuals associated with extreme values of one or more of the independent variables may suggest interactions or cross-product terms (see Chapter 6). Alternately, there may be a grouping of residuals which may correspond to a particular plot, judge, or other unique experimental condition whose effect was not included in the model. Finally, of course, suspect observations may be due to "human errors," such as measurement errors, coding errors, keypunch errors, etc.

4.2 AN EXAMPLE OF SPECIFICATION ERROR

A set of observations have been generated according to the model

$$y = 10 + 2x_1 + x_2 - x_1^2 + \varepsilon$$

where the ε are samples from a normally distributed population with mean zero and unit variance, using a set of 24 arbitrary values of x_1 and x_2. The data are given in Table 4.1. An analysis of data according to the correctly specified model

$$y = \beta_0 + \beta_1 x_1 + \beta_2 x_2 + \beta_3 x_1^2 + \varepsilon$$

provides the estimated equation given in the first column of Table 4.2. The second column gives the estimates obtained when β_3, the coefficient of x_1^2, is omitted from the model.

TABLE 4.1 Example of Specification Error

Obs.	x_1	x_2	y
1	1	1	12.2930
2	1	1	12.7630
3	1	2	13.4250
4	1	2	13.1252
5	2	2	11.4699
6	2	2	12.9083
7	2	2	11.5015
8	2	3	11.3796
9	3	2	9.8865
10	2	3	14.4117
11	2	3	12.9099
12	3	2	9.1078
13	3	3	13.3993
14	3	3	10.4265
15	3	3	9.3504
16	3	3	8.8069
17	4	4	5.5521
18	4	4	4.7936
19	4	2	4.2108
20	4	3	3.6615
21	4	2	5.5764
22	4	2	2.0412
23	4	3	5.5663
24	4	3	4.5436

TABLE 4.2 Summary of Estimation Results

	Correctly specified	Incorrectly specified
F test, H_0: $\underline{\beta} = \underline{0}$	F(3, 20) = 69.34	F(2, 21) = 49.26
R^2	0.912	0.824
Residual MS	1.458	2.782
	Estimated coefficient (standard error)	
β_0	10.009(1.571)	15.867(1.201)
$\beta_1(x_1)$	2.893(1.449)	-3.485(0.374)
$\beta_2(x_2)$	0.681(0.404)	1.206(0.534)
$\beta_3(x_1^2)$	-1.185(0.265)	--

Obviously, knowing the correct model, we see that the results in the second column are "wrong": neither $\hat{\beta}_1$ nor the residual mean square agrees with the true model parameters. However, if the correct model is not known, it is not obvious that a specification error has occurred. The regression as a whole, as well as the estimated coefficients, is statistically significant, the multiple correlation is "respectable," and the residual mean square is relatively small. Without additional information or analysis, these results appear quite acceptable.

A method for detecting specification error is suggested by a further analysis of the residual mean square. An examination of Table 4.1 indicates that there are repeated observations for specific unique combinations of x_1 and x_2. Thus, for example, observations 1 and 2 both have $x_1 = 1$ and $x_2 = 1$, observations 3 and 4 have $x_1 = 1$ and $x_2 = 2$, etc. There exist nine such unique combinations, and the variation of observations within each of these nine sets provides an independent estimate of the error variance, that is, the variation among observations treated alike. This estimate is obtained from an analysis of variance among these nine groups, the results of which are presented in Table 4.3. The residual mean square from this analysis is 1.831, certainly smaller than the residual mean square of 2.782 for the incorrectly specified model. However, an F test for the equality of the two variances is not valid, since these mean squares are not independent. A method does, however, exist for the testing of the significance of this reduction in variance.

The sum of squares for the variation among the groups is associated with eight degrees of freedom, which allows the estimation of *eight* (regression) parameters to describe the variation among the nine groups (for further details see Chapter 7). Subtracting from the between-group sum of squares the sum of squares due to parameters estimated by any model with less than eight parameters provides a sum of squares (and corresponding mean square) contribution of possible additional parameters (see Section 3.4). This is called *lack of fit*, and its expected mean square is σ^2 plus a quadratic form involving the other parameters. The mean square thus provides the numerator for an F test for nonzero values of these additional parameters. This test is shown in Table 4.4 for the (incorrect) model with only two parameters, and it indicates, at the 5% significance level, that the hypothesis of no additional parameters can be rejected [$F(6, 15)_{0.95} = 2.79$]. The same analysis (not reproduced here) involving the lack of fit from the correct model indicates that the hypothesis of no additional parameters is *not* rejected.

Data suitable for this type of analysis do not always exist. However, if other information is available with respect to the magnitude of the error variance, it may be used for a test of the actually obtained mean square. In this example, it is known that the random error has unit variance. This may be used to construct a chi-square test for the hypothesis of unit variance: $\chi^2\ (21) = 21(2.782) = 58.42$, which leads to the rejection of the null hypothesis of unit variance at the 1% level [$\chi^2(21)_{0.99} = 38.9$].

TABLE 4.3 Estimation of Random Error Variance

Source	d.f.	SS	MS	F
Between groups	8	305.060	38.132	20.83
Within groups	15	27.466	1.831	
Total	23	332.526		

TABLE 4.4 Lack-of-Fit Analysis

Source	d.f.	SS	MS	F
Total	23	332.526		
Between groups	8	305.060	38.132	20.83
Due to β_1, β_2	2	274.103	137.051	74.85
Lack of Fit	6	30.957	5.196	2.84
Within groups	15	27.466	1.831	

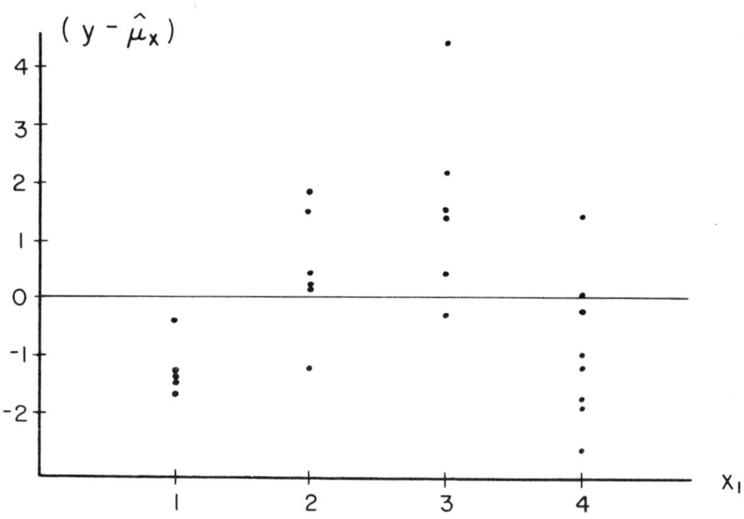

Figure 4.1 Residual plot.

As always, plotting the residuals may help in ascertaining possible deficiencies in the model. Figure 4.1 is a plot of the residuals against the variable x_1. It is easy to see a nonrandom pattern in these residuals: those at the extreme values of x_1 tend to be negative and the values in the middle tend to be positive. Thus, a nonlinear term is suggested; the simplest nonlinear term is the missing quadratic.

4.3 AUTOCORRELATION

The assumption of independently distributed random errors with zero mean and variance σ^2 can be stated in matrix terms:

$$E(\underline{\varepsilon}\underline{\varepsilon}') = \sigma^2 I$$

where E stands for the expected value and the right-hand side represents a scalar multiplication. This simply means that the expected value of any ε^2 is σ^2 and the covariance between any pair ε_i and ε_j, $i \neq j$, is zero. A less restrictive statement on the behavior of the residuals can be written

$$E(\underline{\varepsilon}\underline{\varepsilon}') = V$$

where V is a symmetric positive definite matrix. This definition allows different variances of individual residuals as well as nonzero correlation between pairs of residuals. The corresponding least-squares estimate of β is

$$\hat{\beta} = (X' V^{-1}X)^{-1}X' V^{-1}\underline{y}$$

Other relevant statistics are

$$\text{SS total} = \underline{y}'V^{-1}\underline{y}$$

$$\text{SS regression} = \underline{y}'V^{-1} X(X'V^{-1}X)^{-1}X'V^{-1}\underline{y}$$

$$\text{Variance } (\hat{\beta}) = (X'V^{-1}X)^{-1}\sigma^2$$

This *generalized least-squares estimate* is not directly useful, since the elements of V cannot usually be estimated from available data.

However, there exist certain situations resulting in special forms of the V matrix for which estimation techniques are available.

The most common and well-known situation of this type arises from the analysis of *time series* data, that is, data which are collected continuously over time. In such cases, adjacent observations are often correlated. Similarly correlated errors may also result from other situations, such as manufacturing processes with feedback, observations located along a line in space, etc. This situation is known as *autocorrelation*. We discuss briefly one special case of autocorrelation; more complete discussions are found in books on time series [e.g., Malinvaud (1966)].

A very common form of autocorrelation arises from the *first-order autoregressive model* and occurs as follows. Assume a regression model

$$\underline{y} = X\underline{\beta} + \underline{\varepsilon}$$

intended to describe data collected over successive periods in time, t = 1, 2, ..., n. The $\underline{\varepsilon}$-vector is not the usual vector of independent random residuals, but it arises according to the relationship

$$\varepsilon_t = \rho\varepsilon_{t-1} + \delta_t \qquad (t = 1, 2, \ldots, n)$$

where δ is a true random error with $E(\underline{\delta}\underline{\delta}') = \sigma^2 I$ and ρ is a (usually positive) linear regression coefficient. In other words, the residual of the observation for period t is a function of the residual of the previous period (t - 1) plus a random error, which is the basis for the nomenclature of first-order autoregressive model. The matrix V for the *observed* random errors of the first-order model is

$$V = \sigma^2 \begin{bmatrix} 1 & \rho & \rho^2 & \cdots & \rho^{n-1} \\ \rho & 1 & \rho & \cdots & \rho^{n-2} \\ \cdots & \cdots & \cdots & \cdots & \cdots \\ \rho^{n-1} & \rho^{n-2} & \rho^{n-3} & \cdots & 1 \end{bmatrix}$$

where σ^2 is the common variance of the δ_t.

It is, of course, logical to expect that the use of "ordinary" least-squares methodology for data subject to autocorrelated errors will produce incorrect results and inferences. Johnston (1972, Chapter 8) discusses three aspects of such results:

1. The estimate of β_i
2. The variance of $\hat{\beta}_i$
3. The estimate of σ^2

It is not difficult to show that the $\hat{\beta}_i$ are unbiased estimates of β_i. Unfortunately, general statements on the other aspects of these estimates depend on specific patterns of values of the independent variables. In general, however, both the estimates of the residual variance and the variances of the $\hat{\beta}_i$ are underestimated, with the degree of underestimation being a function of the ratio $(1 - \rho^2)/(1 + \rho^2)$.

The existence of a first-order autoregressive model may be detected by an analysis of the residuals from the model (estimated in the usual fashion) by the use of the Durbin-Watson statistic (Durbin and Watson, 1951):

$$d = \frac{\sum_{t=2}^{n} (\hat{\varepsilon}_t - \hat{\varepsilon}_{t-1})^2}{\sum_{t=1}^{n} \hat{\varepsilon}_t^2}$$

where $\hat{\varepsilon}_t = y_t - \hat{\mu}_{x_t}$. Tables for using d are reproduced in the Appendix. For a one-tail test of positive autocorrelation, the decision rules are

If $d < d_L$, reject H_0: $\rho = 0$

If $d > d_U$, accept H_0: $\rho = 0$

If $d_L \le d \le d_U$, no decision

noting that the tables give 2α as the type I error probability for this one-tail test. For a two-tail test, compute d *and* 4 - d, compare both statistics with the table as above, and use the percentage given for the significance level. For a one-tail test of negative autocorrelation, compute 4 - d only. The fact that some values of d lead to no decision arises from some approximations used in the derivation of the distribution of d.

It has been shown that a close approximation to the generalized least-squares estimate of regression coefficients can be obtained by the use of

$$y_t - \rho y_{t-1}$$

as the dependent variable. Since ρ is not usually known and is difficult to estimate, time series data are often analyzed by using first differences

$$y_t - y_{t-1}$$

which essentially assumes $\rho = 1$, a situation which is not too unrealistic. In many analyses the *independent* variables are also converted to first differences. The resulting coefficients then represent changes in period-to-period differences associated with unit changes in period-to-period differences in the independent variables. In such analyses it is usually necessary to drop the first observation.

Obviously, other types of autocorrelation may occur. The first-order autoregressive scheme generalizes into higher order autoregressive schemes. Cyclic fluctuations may occur in time series data. Various tests for correlated errors have been proposed, for example, the runs test (Ostle and Mensing, 1975, Section 15.4). Other procedures can be found in texts on econometrics or statistics for economics [e.g., Malinvaud (1966)].

A more general methodology for treating such data comes under the general heading of *stochastic processes*, where special techniques are available for discrete or continuous data. Correlated errors not necessarily arising from time series considerations can often be handled by methods of inter- and intra-class correlation [e.g., Ostle and Mensing (1975, Chapter 9)].

4.4 HETEROSKEDASTICITY

The previous section covered the situation where the off-diagonal elements of the generalized error variance matrix (V) were nonzero. Another situation arises where the off-diagonal elements are zero but the diagonal elements are unequal. This situation of unequal error or residual variances is known as *heteroskedasticity*.

Denote the variance of a particular residual as follows:

$$E(\varepsilon_t^2) = \sigma_t^2 \qquad (t = 1, 2, \ldots, n)$$

It can be shown that the corresponding generalized least-squares solution generates a procedure called *weighted regression*, where each observation is weighted by a factor prior to the computation of the sums of squares and cross products. This results in a formula for a particular sum of cross products as follows:

$$\sum_{t=1}^{n} w_t x_{it} x_{jt}$$

where w_t is *proportional* to the reciprocal of the variance σ_t^2. Note that this estimation procedure requires knowledge only of the *relative* variances of the different residuals. Since variance is a measure of lack of precision, weighted regression gives more weight to observations that are more precisely measured or known.

A plot of the residuals from estimated regression values ($\hat{\mu}_x$) may provide information on the type of heteroskedasticity existing in data. In particular, situations where either the variance or standard deviation is proportional to the size of the dependent variable can certainly be detected by noting increasing residuals associated with larger values of the dependent variable. The former occurs in cases where the dependent variable is distributed according to the Poisson or binomial distribution, while the second occurs for variables which tend to vary in proportions or percentages. Much economic data, for example, exhibit this latter property. For such situations, it is not difficult to determine proper weights, such as $1/y$ or $1/\sqrt{y}$, respectively. Alternately, transformations on the observed variables may be employed for these types of heteroskedasticity. However, such transformations are also likely to alter the form of the regression model itself and must be used with caution. The logarithmic transformation is discussed in Chapter 6.

4.5 PROBLEMS

1. Given the following data:

x_1	x_2	y
1	1	4
1	2	5
1	3	9
2	1	6
2	2	13
2	3	17
3	1	8
3	2	18
3	3	27

perform regression analysis using the model

$$y = \beta_0 + \beta_1 x_1 + \beta_2 x_2 + \varepsilon$$

Assume you know $\sigma^2 = 0.67$.

(a) Test for adequacy of model.

(b) Even if part (a) shows that the model is adequate, see if you can detect an inadequacy of specification.

2. The U.S. Environmental Services Administration (Department of Commerce) issues a "Weekly Weather and Crop Bulletin." Among

its contents is a listing of weekly average temperatures for selected cities in the United States. Data for the 14 cities lying near the grid intersections as shown in Figure 4.2 for the first week in January of 1966, 1967, and 1968 are given in Table 4.5. Perform a regression analysis using the model

$$\text{TEMP} = \beta_0 + \beta_1 x_1 + \beta_2 x_2 + \varepsilon$$

where

x_1 = latitude grid code

x_2 = longitude grid code

Perform a statistical analysis to detect specification error(s) and plot residuals. Suggest (but do not implement) alternate methods. An atlas may help.

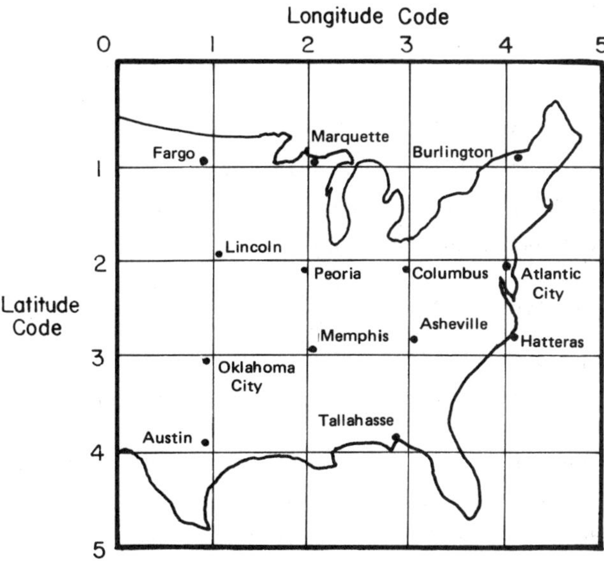

Figure 4.2 Locations of cities.

TABLE 4.5 Temperatures for Week 1

City	Latitude	Longitude	Years		
			66	67	68
Fargo, N. Dak.	1	1	10	9	-5
Marquette, Mich.	1	2	25	22	9
Burlington, Vt.	1	4	30	12	16
Lincoln, Ne.	2	1	34	18	18
Peoria, Ill.	2	2	35	23	16
Columbus, Ohio	2	3	42	29	22
Atlantic City, N.J.	2	4	39	31	27
Oklahoma City, Okla.	3	1	51	31	30
Memphis, Tenn.	3	2	52	37	32
Asheville, N.C.	3	3	42	34	34
Hatteras, N.C.	3	4	54	44	47
Austin, Tex.	4	1	60	45	43
New Orleans, La.	4	2	60	53	46
Tallahassee, Fla.	4	3	60	53	49

3. Quarterly data for 1964-1973 have been collected (Table 4.6) from Business Digest (1976) on possible factors altering consumer sentiment (SENT), a single-valued index (1966 = 100). Perform a regression using as independent variables

   YEAR: a time trend variable (minus 1900)

   MONEY: percent change in the money supply plus time deposits expressed as a compound annual rate

   Compute appropriate test statistics and the predicted and residual values, plot residuals, and compute the Durbin-Watson statistic.

   Redo the model, using first differences of SENT as the dependent variables (keep other variables as they are). Compare results with first analysis.

TABLE 4.6 Data for Problem 3

Obs.	Year	Qtr.	Money	Sent
1	64	1	6.77	99.0
2		2	7.95	98.1
3		3	9.53	100.2
4		4	8.18	99.4
5	65	1	7.92	101.5
6		2	6.95	102.2
7		3	8.53	103.2
8		4	9.40	102.9
9	66	1	6.54	100.0
10		2	5.08	95.7
11		3	3.77	91.2
12		4	3.74	88.3
13	67	1	9.41	92.2
14		2	11.02	94.9
15		3	10.02	96.5
16		4	6.99	92.9
17	68	1	6.72	95.0
18		2	6.49	92.4
19		3	8.35	92.9
20		4	9.44	92.1
21	69	1	5.87	95.1
22		2	3.95	91.6
23		3	0.00	86.4
24		4	1.62	79.7
25	70	1	2.23	78.1
26		2	7.59	75.4
27		3	12.11	77.1
28		4	10.13	75.4
29	71	1	16.15	78.2
30		2	14.35	81.6

TABLE 4.6 (Continued)

Obs.	Year	Qtr.	Money	Sent
31	71	3	9.27	82.4
32		4	10.23	82.2
33	72	1	13.26	87.5
34		2	11.17	89.3
35		3	13.25	94.0
36		4	11.79	90.8
37	73	1	8.54	80.7
38		2	10.24	76.0
39		3	5.22	71.8
40		4	9.72	75.7

4. The temperatures for the first 26 weeks for the city of Austin, Texas (U.S. Department of Commerce, 1966-1968), are given in Table 4.7. Perform a regression using the model

   $$\text{TEMP} = \beta_0 + \beta_1(\text{week}) + \varepsilon$$

   Test for specification error. Specify a better model.

5. Given the following data:

Obs.	x_1	x_2	y
1	1	1	2.0953
2	2	1	3.0533
3	1	2	3.0988
4	1	1	1.6062
5	1	1	2.1526
6	3	1	3.7390
7	1	2	2.7686
8	1	3	5.3812
9	1	1	1.9100

Obs.	x_1	x_2	y
10	3	9	9.9500
11	5	1	7.3940
12	5	2	7.4717
13	5	8	17.0775
14	6	8	17.5389
15	8	1	7.9379
16	9	6	16.7228
17	9	8	23.2831
18	9	8	19.4624
19	8	9	13.3101
20	7	8	15.3107
21	9	9	23.8991
22	2	4	3.7928
23	6	7	18.6401
24	8	2	11.0548
25	7	9	18.8871
26	8	8	19.6620

perform the regression analysis using the model

$$y = \beta_0 + \beta_1 x_1 + \beta_2 x_2 + \varepsilon$$

and plot residuals against $\hat{\mu}_x$ to detect a violation of assumptions. Propose remedy or remedies; implement if possible.

6. It is desired to measure the length of a species of fish as related to age and water temperature. Fish were kept in separate tanks at 25°, 27°, 29°, and 31°C, and a randomly selected specimen was measured from each tank approximately every 14-th day from birth. The data are given in Table 4.8. A linear regression model is proposed:

$$\text{Length} = \beta_0 + \beta_1(\text{Temp.}) + \beta_2(\text{Age}) + \varepsilon$$

TABLE 4.7 Austin Temperatures (Problem 4)

	Year		
Week	1966	1967	1968
1	60	45	43
2	51	45	40
3	52	47	38
4	36	48	49
5	36	59	56
6	47	60	60
7	58	45	49
8	48	54	43
9	43	49	42
10	54	64	52
11	59	61	56
12	66	68	54
13	60	69	55
14	65	75	66
15	66	77	64
16	68	75	68
17	69	77	73
18	73	77	68
19	67	72	70
20	74	83	70
21	80	71	75
22	76	74	78
23	76	79	76
24	81	81	78
25	81	84	84
26	80	86	78

TABLE 4.8 Data for Problem 6

Age	Temp.	Length	Age	Temp.	Length
14	25	620	14	29	590
28	25	1315	28	29	1305
41	25	2120	41	29	2140
55	25	2600	55	29	2890
69	25	3110	69	29	3920
83	25	3535	83	29	3920
97	25	3935	97	29	4515
111	25	4465	111	29	4520
125	25	4530	125	29	4525
139	25	4570	139	29	4565
153	25	4600	153	29	4566
14	27	625	14	31	590
28	27	1215	28	31	1205
41	27	2110	41	31	1915
55	27	2805	55	31	2140
69	27	3255	69	31	2710
83	27	4015	83	31	3020
97	27	4315	97	31	3030
111	27	4495	111	31	3040
125	27	4535	125	31	3180
139	27	4600	139	31	3257
153	27	4600	153	31	3214

Perform appropriate analyses to determine the adequacy of the model (Hint: Use Age × Temp interaction as error term). If inadequacy is detected, suggest a better model (but do not perform the analysis). Note also that since all combinations of AGE and TEMP exist in the data, the SS due to AGE and TEMP are independent and may be used to obtain individually the lack of fit for TEMP and AGE, respectively.

Chapter 5

TOO MANY VARIABLES

5.1 INTRODUCTION

In the discussions of specification error (Chapter 4), it was stated that inadequate specification may result in biased estimates of regression coefficients, while the inclusion of too many parameters will not. For this reason and also because the increasing availability of computers has made easier the estimation of parameters of large linear models, it has become customary to include in a proposed model all parameters that are conceivably relevant, and even some that may be remotely relevant, on the basis that the statistical analysis should eliminate those that are of little value.

However, the inclusion of a large number of parameters often produces large, unwieldy models which are difficult to interpret and use. Such models may also suffer from various side effects of multicollinearity (see Section 5.2). Statisticians have developed a number of tools to help alleviate some of these problems. Foremost among these are variable selection techniques which essentially amount to letting a computer program make the choice of a smaller number of possibly more relevant parameters. Such variable selection procedures are, however, not universally accepted because of problems of statistical inference, mechanics, and optimality.

Before continuing, it may be instructive to give a typical example. This example is concerned with the estimation of the daily amount of evaporation from soil as a function of the temperature of the soil and the surrounding air. Since temperature varies considerably from day to day and hour to hour, it is not certain which aspect of temperature is most important. The following variables may be considered relevant:

1. MAXST = maximum daily soil temperature
2. MINST = minimum daily soil temperature
3. AVST = an index of average daily soil temperature
4. MAXAT = maximum daily air temperature
5. MINAT = minimum daily air temperature
6. AVAT = an index of average daily air temperature

The dependent variable EVAP is the total amount of evaporation from the soil during a given day. Data are available for 25 consecutive days in the month of June and are reproduced in Table 5.1.

The results of standard multiple linear regression analysis relating evaporation to the six temperature variables are summarized in Table 5.2. It is immediately seen that there is a strong relationship between the temperatures and evaporation; the F-statistic easily leads to rejection of the hypothesis of no relationship, and the coefficient of determination (R^2) is 0.839. However, the second portion of the table indicates that only one variable, maximum soil temperature, is statistically significant at the 5% level. In fact, the partial sum of squares for maximum soil temperature is only 326.13, which is a very small portion of the entire regression sum of squares. This is seemingly a rather puzzling and contradictory set of results.

This type of result is, however, quite common when too many variables have been included, since in such situations it is quite common for the independent variables to be highly correlated among themselves. Table 5.3 gives the correlations among all the variables. It is readily apparent that there is "more" correlation among the independent variables than there is between the independent and the dependent variables. Such a phenomenon is known as *multicollinearity*.

One possible, but not necessarily optimum, solution for this problem is to decide to use a smaller number of variables. But which variables? Examining the results in Table 5.2 indicates that possibly

TABLE 5.1 Evaporation Data

DAY	MAXST	MINST	AVST	MAXAT	MINAT	AVAT	EVAP
6	84	65	147	85	59	151	30
7	84	65	149	86	61	159	34
8	79	66	142	83	64	152	33
9	81	67	147	83	65	158	26
10	84	68	167	88	69	180	41
11	74	66	131	77	67	147	4
12	73	66	131	78	69	159	5
13	75	67	134	84	68	159	20
14	84	68	161	89	71	195	31
15	86	72	169	91	76	206	38
16	88	73	178	91	76	208	43
17	90	74	187	94	76	211	47
18	88	72	171	94	75	211	45
19	88	72	171	92	70	201	45
20	81	69	154	87	68	167	11
21	79	68	149	83	68	162	10
22	84	69	160	87	66	173	30
23	84	70	160	87	68	177	29
24	84	70	168	88	70	169	23
25	77	67	147	83	66	170	16
26	87	67	166	92	67	196	37
27	89	69	171	92	72	199	50
28	89	72	180	94	72	204	36
29	93	72	186	92	73	201	54
30	93	74	188	93	72	206	44

TABLE 5.2 Six-variable Regression Analysis

(a) Analysis of variance:

Source	d.f.	SS	MS	F
Total	24	4759.04		
Regression	6	3993.05	665.51	15.639[b]
Residual	18	765.986	42.554	

(b) Coefficients:

Independent var.	$\hat{\beta}$	$t(H_0: \beta_i = 0)$	VIF[c]
Intercept	-164.85		
MAXST	3.1371	2.768[a]	21.96
MINST	-1.4750	-1.064	8.609
AVST	-0.4067	-0.985	27.93
MAXAT	0.4073	0.392	14.40
MINAT	0.7042	0.734	10.09
AVAT	0.0869	0.340	17.81

[a]Statistically significant at 0.05 level.
[b]Statistically significant at 0.01 level.
[c]VIF is the variance inflation factor to be explained later.

TABLE 5.3 Correlations

	MAXST	MINST	AVST	MAXAT	MINAT	AVAT	EVAP
MAXST	1.00	0.75	0.95	0.93	0.50	0.83	0.89
MINST	0.75	1.00	0.87	0.78	0.85	0.85	0.61
AVST	0.95	0.87	1.00	0.93	0.68	0.89	0.82
MAXAT	0.93	0.78	0.93	1.00	0.63	0.91	0.85
MINAT	0.50	0.85	0.68	0.63	1.00	0.83	0.45
AVAT	0.83	0.85	0.89	0.91	0.83	1.00	0.77

maximum soil temperature should be selected. A two-step analysis providing the sum of squares due to maximum soil temperature only, and the additional sum of squares due to all other variables, is provided in Table 5.4. Maximum soil temperature is found to be highly significant, while the inclusion of all the other variables actually *increases* the residual mean square, when ordinarily a reduction would be expected.

The use of the statistical significance tests of the individual partial coefficients in a regression containing too many variables is not usually, however, an optimum procedure for variable selection. Procedures for performing this task and other techniques which may be useful in the evaluation of the importance of variables are given in Section 5.3.

TABLE 5.4 Two-Step Analysis

Source	d.f.	SS	MS	F
Total	24	4759.04		
MAXST (only)	1	3797.68	3797.68	90.858[a]
Residual	23	961.357	41.798	
Other variables (adj)	5	195.371	39.074	0.918
Residual	18	765.986	42.554	

[a]Statistically significant at the 0.01 level.

5.2 MULTICOLLINEARITY

The problem of multicollinearity has plagued users of regression analysis for many years, even before the advent of computers made this phenomenon more prevalent by seemingly inviting users to dump into the model an unlimited number of variables. In many situations it is important to the researcher to try to separate the relative effects on the dependent variable of a number of interrelated, independent variables. Examples of this type of problem include

1. Relating household consumption (y) to income (x_1) and education (x_2); x_1 and x_2 are correlated because income is usually related to education
2. Relating retail sales (y) to per capita income (x_1), price level (x_2), and employment (x_3) using data over a period of time; if per capita income *and* price levels are based on *real* dollars, they will usually be related by the existence of inflation
3. Relating yield of corn (y) to height of stalk (x_1) and diameter of stalk at base (x_2); both x_1 and x_2 are growth measurements and will thus be related
4. Relating quality of performance (y) of a task by individuals to their IQ (x_1) and education (x_2); persons with higher IQs tend to have more education; hence, these measurements will be related
5. In estimating polynomial relationships, it often happens that x and its powers are highly correlated, particularly if the range of x (in the data) is small relative to the mean of x.

The existence of multicollinearity does not invalidate a regression analysis. The absence of multicollinearity is not one of the underlying assumptions (see Section 2.3 and Chapter 4). As long as the independent variables are measured without error, the analysis is valid. Multicollinearity is also not a specification error. However, as seen in the example of the previous section, multicollinearity does introduce some rather serious problems. These can be summarized according to the various aspects associated with regression analysis, as follows.

1. *Estimation.* The estimating ability of the regression equation, that is, the estimation of the degree of relationship of $\hat{\mu}_x$ to y, is *not* affected by multicollinearity. In other words, regardless of the existence of multicollinearity, the residual mean square is a

valid statistic for the ability of the regression model to estimate the population mean for a specific set of x_i.

2. *Prediction*. It is often required to predict the response for a specific set of x-values not necessarily corresponding to a sample data point. In other words, using the first example above, it may be desired to predict mean consumption for individuals with low income and high level of education, although there were no such individuals in the sample. Such predictions are usually extremely unreliable, since they are outside the range of the data, even though not necessarily out of range of the individual independent variables. They should be considered in the same light as extrapolations.

3. *Computations*. Multicollinearity is a high degree of linear relationship among independent variables. An *exact* linear relationship among two or more independent variables creates a singular X'X matrix which cannot be inverted. A high degree of multicollinearity produces a "near singular" matrix, which is subject to severe round-off errors in most matrix inversion algorithms.

4. *Interpretation*. An important aspect of most regression analyses lies in the interpretation of individual partial regression coefficients. Actually, the use of a partial regression coefficient in the presence of multicollinearity presents somewhat of a contradiction. Remember that a partial regression coefficient shows the change in the dependent variable associated with a unit change in the corresponding independent variable, *all other variables remaining constant*. However, the existence of multicollinearity implies that the independent variables are very strongly related to each other; thus it is, by definition, nearly impossible for one variable to vary while all other variables remain at constant levels. In a sense, a partial regression coefficient under multicollinearity represents an illogical extrapolation from nonexistent data. It is, therefore, comforting that the statistical analysis of partial regression coefficients reflects this dilemma by indicating instability (large variance) of the estimated regression coefficients.

The variance of $\hat{\beta}_i$ has been shown to be $c_{ii}s^2$. The usual estimate of σ^2 is not affected by multicollinearity (see above). It can be shown that

$$c_{ii} = \frac{1}{(1 - R_i^2)\sum(x_i - \bar{x}_i)^2} \tag{5.1}$$

where $\Sigma(x_i - \bar{x}_i)^2$ is the sum of squares of the corresponding independent variable, and R_i is the multiple correlation between x_i and all other independent variables, that is, R_i is a measure of multicollinearity affecting x_i. Since the quantity $\Sigma(x_i - \bar{x}_i)^2$ is not affected by multicollinearity, a large R_i^2 will increase c_{ii}, hence, also the variance of $\hat{\beta}_i$. Specifically, the variance is "inflated" by the quantity $1/(1 - R_i^2)$, which may be called the *variance inflation factor* (VIF) (R. D. Snee, 1973). This factor is easily computed: it is the product of the corresponding elements of the corrected X'X and C matrices,

$$\text{VIF} = c_{ii}[\sum(x_i - \bar{x}_i)^2] \tag{5.2}$$

VIF values for the example of the previous section are given in the last column of the coefficient listing of Table 5.2. Obviously, multicollinearity is very badly affecting the precision of the coefficient estimates as the variances are inflated by factors as high as 28.

Having now discussed the existence of, and problems associated with, multicollinearity, we need to consider possible methods to avoid it or minimize its effect. Unfortunately there is no simple or clear-cut solution. Some recommendations can, however, be made:

1. Care in the initial choice of variables is important. If it is known that several variables essentially measure the same factor, then more than one of these should not be used without good reason
2. If prior information cannot be used to limit the number of variables, a variable selection procedure (Section 5.3) may

be used. Unfortunately, such procedures are less effective in the presence of multicollinearity

3. Occasionally, certain transformations or recombinations may help. For example, x and x^2 may be highly correlated but $x - \bar{x}$ and $(x - \bar{x})^2$ are uncorrelated. In some situations, deflation by the use of a factor common to a number of variables will decrease multicollinearity, e.g., price levels when several prices are used or size of an organism when several size measurements have been taken
4. Certain multivariate techniques such as principal components or factor analysis [see, e.g., Morrison (1976)] may sometimes be used to help in the selection of independent variables.
5. In some situations, theory or other considerations may dictate that correlated variables must be included in the model. For such situations, biased estimation procedures such as *ridge regression* may provide usable estimates of regression coefficients (see Section 5.5).

5.3 VARIABLE SELECTION

In the absence of a properly specified model, it is common practice to consider for inclusion a large number of variables, or functions of variables (and associated parameters), then to select a subset of variables or terms which appear most relevant, and finally to specify the model on the basis of such selected terms. This does not constitute a strictly valid statistical procedure, since the correctness of the type I and II error probabilities associated with statistical hypothesis tests assumes that a single model and a restricted set of hypotheses have been specified *prior* to data collection. In other words, the model and associated hypotheses should *not* be determined by the data. Obviously, the very act of variable selection violates this assumption; hence, the stated significance levels of the associated tests are incorrect (see Section 5.6).

However, the initial inclusion of a large number of variables may be justified by the fact that the omission of essential variables may produce biased estimates (see Chapter 4) while the inclusion of extraneous variables does not. Furthermore, much research is of an essentially exploratory nature, thus precluding correctly specified models. For these reasons, statisticians have developed a number of variable selection procedures.

There are two questions which a variable selection procedure tries to answer:

1. How many variables should be included?
2. Given the number of variables to be included, what specific subset of variables provides the minimum residual mean square, the usual criterion of optimum selection?

These aspects are highly related and most variable selection procedures combine them.

It is usually necessary to examine a large number of variable subsets in order to find the optimum subsets (Section 5.3.6). This involves considerable computation and quickly becomes expensive even for large computers. For this reason, the majority of practical applications of variable selection are implemented by simpler procedures which, although not guaranteeing optimal subsets, generally provide acceptable results.

Three procedures which have been widely used are

1. Backward elimination (step-down)
2. Forward selection (step-up)
3. Stepwise

Following a discussion of these, we present a brief discussion on other procedures currently in use. Computational aspects of all procedures are covered briefly, since practical considerations necessitate the use of computers and well-documented programs implementing most of these procedures are generally available.

The choice of a selection procedure to be used for any specific problem is not completely clear; it often depends on the availability of computational facilities and appropriate computer programs, the degree of optimality required, and the personal inclinations of the user.

5.3.1 Backward Elimination

In this procedure, an initial analysis is performed using all variables. The t-statistic for significance of each coefficient is calculated and the regression coefficient with minimum absolute t-value, i.e., the "least" significant coefficient, is deleted. The remaining variables produce an equation with one less coefficient. All t-statistics are computed for the new regression and again the least significant coefficient is deleted, etc. This procedure is a one-at-a-time *elimination*, or a *step-down* procedure. The final model may be selected according to various criteria: when all remaining coefficients are statistically significant at some predetermined level, by examining all steps and picking an optimum point (see C_p plots, Section 5.3.7), or other criteria.

It is important to note that once a variable has been deleted, it may not re-enter. This feature may prevent the finding of an optimum combination of variables, since a deleted variable may, once another variable has been deleted, become quite important. Such phenomena occur particularly often when there is a high degree of multicollinearity.

The mechanics of this procedure are simple. The least significant coefficient corresponds to the smallest (absolute) t-statistic, hence also the smallest value of $\hat{\beta}_i^2/c_{ii}$. The regression with that variable deleted may be computed by the elimination procedure given in Section 1.6.

5.3.2 Forward Selection

In contrast to backward elimination, this procedure starts with the "best" one-variable equation and adds additional variables, one at a time. Hence, it is sometimes called a *step-up* procedure. The independent variable for the best single-variable equation is obtained by seeking the maximum of the (absolute) pairwise correlations of the independent variables and the dependent variable. The second variable is chosen by seeking the maximum of the partial correlations of the remaining independent variables with the dependent variable, adjusted for the variable already chosen. The sum of squares due to the addition of variables can be calculated and corresponding statistical significance tests obtained. As in backward elimination, the procedure is continued until some criterion is met: further additions are not statistically or practically significant, etc. Again, the unidirectional aspect, by which, in this case, selected variables may not be dropped, can lead to nonoptimal selection.

Computationally, this procedure has many of the aspects of the step-down procedure. The key to the computations lies in the recursion formulas for the partial correlation coefficient given in Chapter 3 [formulas (3.28) and (3.29)].

5.3.3 Stepwise

This procedure is a refinement of the forward selection procedure. At each step, before the determination of the next variable to be *added*, the statistics for significance of the already chosen coefficients are examined to see if a variable *elimination* may be in order. This procedure can thus be described as a step-up procedure with a step-down adjustment. The procedure stops when, for specified significance levels, neither a forward selection nor a backward elimination is indicated. Intuitively, this procedure would seem to overcome the major objections to either step-up or step-down, but it does *not* guarantee the choice of optimum subsets.

The computational procedures for this method are, of course, more difficult than for either the step-down or step-up procedures. A convenient procedure in tableau form is available for hand-calculator usage (Draper and Smith, (1966)).

5.3.4 Discussion

It has already been indicated that these selection procedures do not guarantee optimum subsets and that other methods exist which do, at greater cost, guarantee an optimum subset of variables. However, the above procedures have been used extensively in the past and their relative computational simplicity will cause them to continue to be used.

The backward elimination procedure has great intuitive and statistical appeal, since it starts with the entire model and thus cannot be accused of overlooking anything. The first step naturally does provide the optimum subset, and generally the earlier steps, those with a relatively large number of variables, provide more nearly optimum subsets than the other procedures, since fewer potential candidates for re-entry will have been deleted. On the other hand, if there exist linear dependencies or near dependencies, the initial model with all variables may exhibit the difficulties listed under multicollinearity.

The forward selection procedure is computationally the easiest. The one-variable equation is naturally optimum, and generally the equations with a small number of variables will provide more nearly optimum subsets, since fewer candidates for deletion will have been chosen. Also, multicollinearity or exact linear dependencies do not affect this procedure, since the entire $X'X$ matrix (or equivalent) is never inverted. A disadvantage is that, unless all variables are forced into the equation, the effectiveness of the equation containing all variables cannot be investigated.

The stepwise procedure overcomes some of the major deficiencies encountered in the step-up and step-down methods and is probably the best of these three procedures, but is still not optimum. There is, in fact, considerable doubt that the additional computational effort is worthwhile. Also, the final results of the procedure are not unique, as they are dependent on choices of significance levels for forward selection and backward elimination.

5.3.5 An Example

Difficulties with selection procedures arise primarily in problems with many independent variables which, by the sheer volume of data, are unsuitable as text examples. The selection procedures are therefore illustrated here by the use of a small artificial data set. The data are given in Table 5.5 and the $X'X$ and $X'\underline{y}$ matrices (note that all means are zero) in Table 5.6. (A more comprehensive example is given in Section 5.3.8.)

In order to illustrate more clearly the principles and properties of the procedures (rather than the mechanics of calculation), Table 5.7 provides summary statistics for the inclusion of *all possible* combinations of variables. This allows the reconstruction of the various procedures.

The step-down approach proceeds as follows:

Step 1. In the full equation, $\hat{\beta}_3$ has the smallest t-value; hence, we delete x_3, leaving (x_1, x_2, x_4) with a residual mean square of 54.0.

Step 2. In the equation using (x_1, x_2, x_4), $\hat{\beta}_2$ has the smallest t-value; hence, the choice is (x_1, x_4) with a residual mean square of 88.8.

Step 3. In this equation, $\hat{\beta}_1$ has the smaller t-value; hence, the better equation uses x_4 with a residual mean square of 552.5.

The step-up procedure proceeds as follows:

Step 1. The optimum one-variable regression (with smallest MSE) involves x_3; the residual mean square is 347.0.

Step 2. The optimum two-variable regression which includes the already chosen x_3 involves (x_1, x_3); the residual mean square is 164.3.

Step 3. Having chosen x_1 and x_3, the choice is restricted to (x_1, x_2, x_3) or (x_1, x_3, x_4); of these, (x_1, x_3, x_4) is optimum with residual mean square of 55.1.

TABLE 5.5 Data for Section 5.3.5

Observation	x_1	x_2	x_3	x_4	y
1	42	14	64	71	57
2	27	69	-13	0	-32
3	48	0	-7	0	-32
4	0	0	27	0	9
5	0	0	0	0	9
6	-42	-14	-64	-71	-57
7	-27	-69	13	0	32
8	-48	0	7	0	32
9	0	0	-27	0	-9
10	0	0	0	0	-9

TABLE 5.6 X'X Matrix

	x_1	x_2	x_3	x_4	y
x_1	9594	4902	4002	5964	-12
x_2	4902	9914	-2	1988	-2820
x_3	4002	-2	10086	9088	9062
x_4	5964	1988	9088	10082	8094
y	-12	-2820	9062	8094	10918

TABLE 5.7 Summary of All Variable Combinations

Vars. in Eqs.	$\hat{\beta}_1$ / t_1	$\hat{\beta}_2$ / t_2	$\hat{\beta}_3$ / t_3	$\hat{\beta}_4$ / t_4	SSE	MSE	R^2
Mean					10918	1213.1	0.0000
1	-0.0013 -0.0033				10918	1364.8	0.0000
2		-0.2844 -0.7965			10116	1264.5	0.0735
3			0.8985 4.8439		2776	347.0	0.7457
4				0.8028 3.4294	4420	552.5	0.5951
1, 2	0.1928 0.4352	-0.3798 -0.8715			9849	1407.0	0.0979
1, 3	-0.4506 -3.1453		1.0773 7.7096		1150	164.3	0.8946
1, 4	-0.7913 -6.5391			1.2709 10.7664	621	88.8	0.9430
2, 3		-0.2843 -1.6851	0.8984 5.3717		1975	282.13	0.8191
2, 4		-0.4638 -2.4584		0.8943 4.7805	2372	338.9	0.7827
3, 4			0.9324 2.0387	-0.0377 -0.0824	2773	396.2	0.7460
1, 2, 3	-0.3967 -2.1920	-0.0881 -0.5417	1.0558 6.9199		1097	182.8	0.8995
1, 2, 4	-0.6667 -6.1583	-0.2029 -2.3483		1.2372 13.2813	342	54.0	0.9703
1, 3, 4	-0.6942 -6.6571		0.4288 2.2981	0.8269 3.8563	331	55.1	0.9697
2, 3, 4		-0.3507 -1.7851	0.6003 1.3627	0.3309 0.7861	1811	301.9	0.8341
1, 2, 3, 4	-0.6185 -7.1348	-0.1591 -2.2818	0.3333 2.2360	0.8993 5.3710	162	32.4	0.9852

The stepwise procedure would follow steps 1, 2, and 3 of step-up. Assuming the α level for elimination at 0.05, the elimination of x_3 would be nonsignificant; hence, the next regression is the two-variable combination of x_1 and x_4. Next (depending on the significance level for forward selection *add* x_2 (since both remaining coefficients are statistically significant) to obtain the three-variable equation using x_1, x_2, and x_4 with a residual mean square of 54.0. Depending again on the specification of stopping rules, the procedure would then either add x_3 or stop altogether. In summary, this procedure would select subsets as follows: (x_3), (x_1, x_3), (x_1, x_3, x_4), (x_1, x_4), (x_1, x_2, x_4).

Note that in this example, backward elimination is superior for selecting a two-variable equation while forward selection is superior for the best single variable. The stepwise procedure finds all optimum combinations. A graphical method for evaluating variable selection which is particularly useful for large numbers of variables is discussed in Section 5.3.7.

5.3.6 Other Selection Procedures

The fact that the selection procedures outlined in Section 5.3.5 do not guarantee to select optimum subsets has caused considerable effort to be expended in developing procedures that do guarantee such subsets. One approach has consisted of developing efficient algorithms for examining all possible subsets. Such algorithms, aided by efficient computer programming, make this approach at least manageable for problems of up to 15 or 20 independent variables.

Another approach has consisted of the development of more effective search procedures that reduce the total number of subsets to be considered. A variable selection procedure can be compared to a 2^n factorial experiment for which various experimental designs, such as fractional factorial designs, can be used to estimate regions of minimum residual mean squares. Another development by LaMotte and Hocking (1970), called "SELECT," uses a theorem of least squares to provide a test of whether a number of subsets, selected in a certain way, contain the optimum subset. This procedure often reduces by a factor of several thousand the number of subsets to be examined to insure an optimum. It is still, however, quite costly for more than 30 independent variables. Details of procedures such as these are not discussed here, since they involve relatively complex mathematical arguments and require sophisticated computer programs for implementation.

5.3.7 Evaluation of Selected Subsets

The "step" procedures normally provide only one equation for each subset size, and significance tests are often used to determine the appropriate size of subset. In some situations, particularly when sample sizes are large, statistically significant coefficients may provide practically insignificant contributions to the fit of the equation. A plot showing, for example, how the residual mean square increases as variables are deleted may provide a better indicator of appropriate subset size.

For procedures which examine a large number of subsets, statistical significance tests are meaningless, since correct type I error probabilities are essentially unobtainable. Such procedures also provide a number of "nearly" optimum subsets, which may be more appealing than the optimum subset, since they may use variables that may be easier to measure, more meaningful in a practical sense, or exhibit less multicollinearity, hence providing better interpretation of the data. Studies of such subsets may also help to reveal underlying structure which can be valuable in selecting useful subsets.

If a large number of near-optimal regression subsets have been computed, it is instructive to summarize the results in a plot. One such plot is the C_p plot proposed by Mallows (1973). This consists of plotting for subsets of size p, say, the quantity

$$C_p = \frac{SSE_p}{\hat{\sigma}^2} - (n - 2p - 1) \tag{5.3}$$

where

SSE_p = residual sum of squares for the p selected variables

$\hat{\sigma}^2$ = estimate of the error variance

n = total sample size†

Using the residual mean square from the fitting of all m independent variables as $\hat{\sigma}^2$, it is easy to show that

$$C_p = (m - p)F - (m - 2p) \tag{5.4}$$

where F is the statistic for testing the null hypothesis that all coefficients beyond the p already selected are zero. When F equals 1, that is, when the residual mean square is the same for all m or the selected p variables, then $C_p = p$; when F is less than 1, that is, when the residual mean square using p selected variables is *less* than when all variables have been used, then C_p is less than p, and vice versa. When any $C_p < p$ (or $F < 1$) the original (m-variable) equation is considered to contain too many variables, since the residual mean square is actually reduced by deleting variables; such an equation is said to be overspecified.

It is instructive to plot values of C_p against p for the optimum and near-optimum subsets. In such a plot, the existence of an overspecified equation is made evident by points below the 45° line $C_p = p$. In such cases, it is usually recommended that the number of variables selected lie between the minimum C_p and the point where C_p crosses the 45° line. An example of such a plot is given in Section 5.3.8.

It is also instructive to examine some nearly optimum subsets for clues on interrelationships among the independent variables. The choice of the most appropriate subset(s) is somewhat subjective and may depend on nonstatistical considerations.

---

†The equation given in Mallow's paper is slightly different due to the fact that he includes β_0 as one of the coefficients in the selection process.

An equivalent formula for this statistic is

$$C_p = (n - m - 1) \frac{1 - R_p^2}{1 - R_m^2} - n + 2p + 1 \tag{5.5}$$

where R_p and R_m are the multiple correlations for the p subset and all m variables included, respectively. Obviously, C_p is only one of many equivalent statistics which may be plotted as an aid to variable selection.

5.3.8 An Example

In grading meat it is desirable to use easily recognizable criteria as indicators of meat quality. An experiment was conducted using rib sections from 403 pork carcasses and the following four predictive criteria:

A. An index of marbling (thin fatty streaks within the main body of meat), coded from 1 to 5, corresponding to increasing marbling

B. An index of color, coded from 1 to 5, corresponding from light to dark red

C. An index of firmness, coded from 1 to 5, corresponding to soft to firm flesh

D. Gross morphology, a combination of color and firmness, with 1 corresponding to soft and light to 5 being firm and dark

The quality characteristic of meat to be predicted from these is *juiciness*, which is an objectively determined measure ranging from about 2.5 to 7.5 in increasing degrees of juiciness.

The 403 ribs were taken as they were available and thus do not correspond to a design with all possible combinations of A, B, C, and D. In fact, as expected, only 183 of a possible 625 of these combinations were available. There were, however, many combinations for which there existed a number of carcasses which could be used as a basis for estimating experimental error. An initial exploratory model was proposed which included coefficients for the following functions of the four independent variables: A, A^2, B, B^2, C, C^2, D, D^2, AB, AC, AD, BC, BD, CD. An interpretation of this type of model is given in Chapter 6. The lack-of-fit analysis is given in Table 5.8. The lack of fit is not statistically significant, although it is conceivable that a few individual terms not used in the 14-variable model might produce a residual mean square closer to the true error mean square of 0.608. Residual plots do not, however, reveal any obvious misspecifications.

The individual coefficients and associated significance tests and variance inflation factors are given in Table 5.9. The model is obviously overspecified; the original equation is highly significant, while none of the individual coefficients indicate such significance. The variance inflation factors are very large, implying a large degree of multicollinearity. A variable selection procedure is indicated.

TABLE 5.8 Initial Equation Analysis

Source	d.f.	SS	MS	F
Total	402	371.69		
Between combinations	182	237.90		
Due to 14 variables	14	106.01	7.572	12.45[a]
Lack of fit	168	131.89	0.785	1.29
Error	220	133.79	0.608	

[a]Statistically significant at the 0.01 level.

TABLE 5.9 Fourteen-Variable Equation

Coefficient for	Est. $\hat{\beta}_i$	t for H_0: $\beta_i = 0$	VIF
A	0.2033	1.15	39.4
A^2	-0.0007	-0.02	54.3
B	0.3575	1.06	73.4
B^2	0.1124	1.14	114.1
AB	0.0534	0.64	46.0
C	0.2100	0.63	116.8
C^2	0.0846	0.77	243.7
AC	0.0840	1.14	102.4
BC	0.0073	0.04	118.0
D	0.2701	0.54	137.2
D^2	0.3791	1.33	304.4
AD	-0.1387	-1.10	313.7
BD	-0.3638	-1.30	863.5
CD	-0.3247	-1.04	1370.6

An initial step-down analysis indicated that at most five variables would be needed. In order to obtain the best equations, all possible combinations were examined for one to five variables, and a partitioning of sums of squares indicating the progressive reduction in the residual sum of squares was performed to indicate a reasonable stopping point. This analysis is presented in Table 5.10, and the rapid decrease in the F-statistic clearly indicates that four variables are adequate. It should be noted that since all possible equations were examined, the F-statistic should only be used as a guide rather than a true indicator of statistical significance.

TABLE 5.10 Reduction of Residuals by Adding Variables

Source	d.f.	SS	MS	F
Total	402	371.69		
Best variable	1	80.92	80.92	111.60
residual (1)	401	290.77	0.7251	
Add second variable	1	11.05	11.05	15.79
residual (2)	400	279.72	0.6998	
Add third variable	1	4.84	4.84	7.03
residual (3)	399	274.87	0.6889	
Add fourth variable	1	6.00	6.00	8.89
residual (4)	398	268.87	0.6756	
Add fifth variable	1	0.68	0.68	1.01
residual (5)	397	268.18	0.6755	
All other variables	9	2.50	0.28	0.41
residual (14)	388	265.68	0.6848	

The 10 most nearly optimum equations for one through four variables are summarized in Table 5.11. The results of the forward selection (step-up) and backward elimination (step-down) are also indicated in this table; the rank of optimality of the chosen equation is given in parentheses. The stepwise procedure was also implemented with a deletion significance level of 0.10 and a forward selection significance level of 0.50, which resulted in the four variable equation given in the last line of Table 5.10. It is obvious that the step-type analyses have not performed particularly well here, although the coefficients of determination are certainly not much lower than they are for the best equations. It is quite surprising that the backward elimination does find the optimum two-variable equation.

The information in Table 5.11 is summarized in the C_p plot given in Figure 5.1. The residual mean square for the 14-variable equation was used as the estimate of variance; the use of the error mean square from the lack-of-fit analysis would have provided points

TABLE 5.11 Summary of Selection Procedures

Selection	No. variables included							
	1		2		3		4	
	Var.	R^2	Var.	R^2	Var.	R^2	Var.	R^2
Best	AD	0.218	A B	0.247	B AC BC	0.260	A B C BC	0.276
2	AC	0.217	B AC	0.244	A B AB	0.259	B C AC BC	0.275
3	AB	0.212	A D	0.243	B AC BD	0.257	A^2 B C BC	0.274
4	A	0.208	A^2 B	0.243	A D D^2	0.256	A B C^2 BC	0.274
5	A^2	0.195	A^2 D	0.240	A B C	0.256	A B^2 D BD	0.272
6	D	0.168	A B^2	0.239	A^2 D D^2	0.255	A B^2 C BC	0.272
7	C	0.165	A BC	0.236	A B C^2	0.254	B AB C BC	0.271
8	CD	0.156	A BD	0.236	B B^2 AC	0.254	A^2 B^2 D BD	0.271
9	BC	0.153	A C	0.236	A B AC	0.253	A^2 B^2 C BC	0.269
10	C^2	0.151	B^2 AC	0.236	A^2 B C	0.252	A^2 B C^2 BC	0.269
Step-up (no.)	(1)		(19)		(29)		(42)	
	AD	0.218	B AD	0.230	A B AD	0.247	A B AB AD	0.263
Step-down (no.)	(4)		(1)		(25)		(22)	
	A	0.208	A B	0.247	A B BD	0.248	A B D^2 BD	0.265
Stepwise (no.)							(12)	
							A B AB AC	0.268

above the 45° line, but the configuration of points would have been the same. The four-variable equation again seems to be a best choice, but not clearly so.

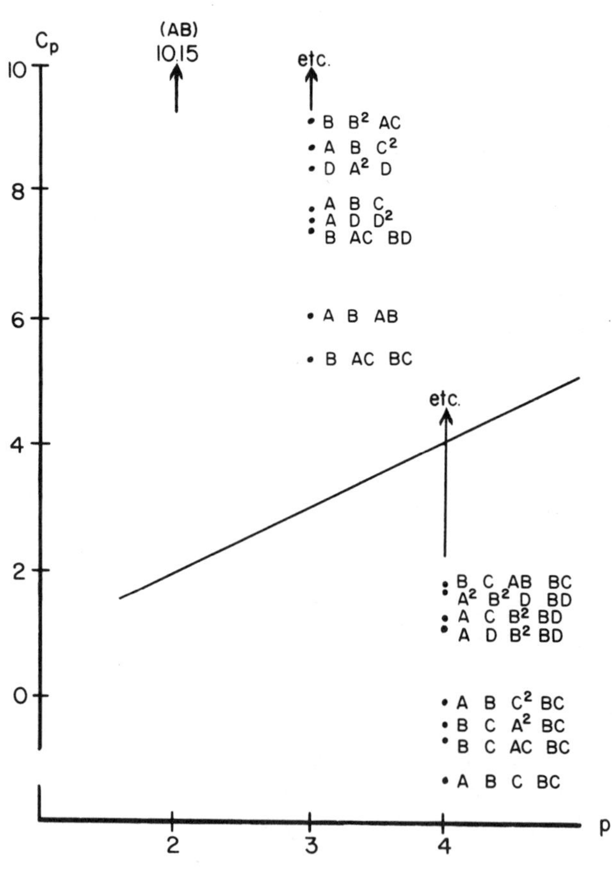

Figure 5.1 C_p plot data.

On examining these equations it is rather obvious that variables A, B, and C together with the BC interaction provide the best description of the relationship between the observed measurements and juiciness. Note that the inclusion of the gross morphology (D) provides equations with higher C_p than do those not using this factor. The various equations do not, however, differ very markedly, and so an ultimate choice would probably involve a subjective judgment on the usefulness of these equations. Note, for example, that the second best of three-variable equations involves only A and B, which, needing only two measurements, is an equation much easier to implement in practice than one which includes A, B, and C.

5.4 OTHER PROCEDURES

The dramatic reduction of the cost of computations in recent years has spurred considerable interest in the subject of how to deal with regressions containing too many variables. This topic is the subject of much current research, and many papers on this topic have been published in recent issues of the journal Technometrics. Much of this research is highly mathematical in nature and is beyond the scope of this book, and little of the proposed methodology has been fully evaluated or implemented in the widely available statistical computing packages.

Most of the work in this subject can be divided into two categories:

1. More efficient and/or effective methodologies for variable selection
2. Methods which do not necessarily result in the deletion of variables from a regression equation

We have already briefly discussed some of the newer methods of variable selection and the evaluation of selected subsets. Alternate methods revolve primarily around biased estimation, that is, procedures which produce estimates of the regression coefficients which are biased but may, on the other hand, have smaller variances and therefore still be quite useful. A number of biased estimation procedures have been proposed and the most popular of these (*ridge regression*) is discussed below.

5.5 AN EXAMPLE OF RIDGE REGRESSION

5.5.1 Introduction

The least-squares estimation of parameters in a linear (regression) model is known to produce unbiased estimates. There are, however, estimating procedures which produce estimates that are not necessarily unbiased; that is, over all possible samples, these estimates do not average to the population value of the parameter. Such biased estimators may, however, have a very small variance. Hence, they may provide estimates that are more reliable; that is, on the average, they may be nearer the population parameter value than the minimum variance unbiased estimate.

The concept of a biased estimator of a parameter is illustrated in Figure 5.2, which gives hypothetical sampling distributions of two estimates of a population parameter, θ. One of the estimates is unbiased; that is, the mean of the sampling distribution is indeed the population parameter. However, the distribution has a large variance. The other procedure is biased, as it almost always provides sample estimates that are larger than the population parameter values; however, the variance of the estimate is quite small. Obviously, the choice of which of these two estimates to use is not completely clear. The biased estimate would certainly be the choice if there were heavy penalties for large errors.

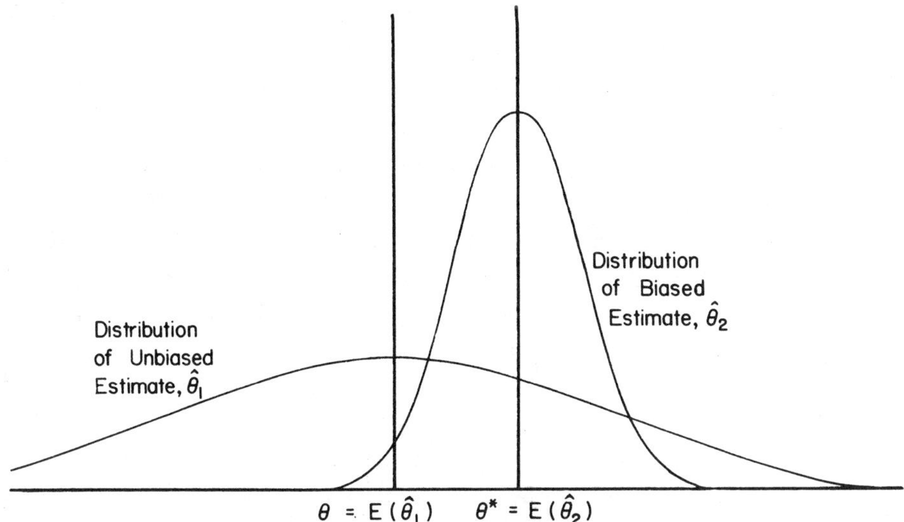

Figure 5.2 Sampling distribution of estimates.

An expression which provides an index of the relative quality or reliability of an estimate is the *mean square error*, which is the sum of the variance and the square of the bias of the estimate; the bias in the case of the biased estimate in Figure 5.2 would be $\theta^* - \theta$. In the example of Figure 5.2, it is quite likely that the mean squared error of the biased estimate is smaller than that of the unbiased estimate.

Statisticians tend to avoid biased estimates. This is largely due to a reluctance to use estimates which on the average miss the mark. Other reasons for not using biased estimates is that the bias is often not a decreasing function of sample size and that the magnitude of bias is often difficult or impossible to estimate. Therefore, one cannot always be certain that a given biased estimate does, in fact, have a smaller mean square error than a corresponding unbiased estimate.

Ridge regression (Alldredge and Gilb, 1976) is a procedure for obtaining biased estimates of regression coefficients which is particularly useful for situations where the variances of the regression coefficients are inflated by the existence of multicollinearity. In ridge regression a reduction in the variance inflation factor is "bought" by introducing a bias; if multicollinearity is extreme, one can be rather certain that the mean square error of the resulting regression coefficient estimates is reduced.

The least-squares estimate of a vector of regression coefficients as given in Chapter 2 is

$$\hat{\underline{\beta}} = (X'X)^{-1}X'\underline{y} \tag{5.6}$$

Assume that all variables have been reduced to equal variance; hence, X'X is a correlation matrix and has values of 1 on the diagonal, and $\hat{\underline{\beta}}$ is the vector of standardized coefficients. The ridge regression estimate is provided by the following formula:

$$\underline{\beta}^* = (X'X + kI)^{-1}X'\underline{y} \tag{5.7}$$

where kI represents a scalar multiplication of the identity matrix by a small positive number k. In other words, a small positive constant has been added to each element on the diagonal of the correlation matrix. This *automatically* reduces variance inflation factors, since the resulting individual pairwise "correlations" have larger denominators.

A problem now arises: what should be the value of k? The most frequently proposed method is to provide ridge regression estimates for a sequence of values of k, plot the resulting coefficients against k (this is known as a *ridge plot*), and pick the k-value where the plot seems to "settle down." This procedure is subjective and the chosen k may depend on the scale used for k. However, results apparently are not badly affected by an inexact choice of k, and there is continuing research on more optimum choices for k. Since no universally acceptable method has yet been derived, the example used here will use the subjective choice of the k-value from a ridge plot.

5.5.2 An Example

The data concern the estimation of the total production of biomass of mesquite using easily measured parameters of the plant as opposed to the actual harvesting of mesquite. Two separate data collections are available, one labeled MCD, consisting of measurements on 26 mesquite bushes, and another labeled ALS, containing measurements on 20 mesquite bushes. Both data sets were obtained in the same geographic location (ranch) but were taken at different times of year, and neither constitutes a strictly random sample. The data are given in Table 5.12; the data labeled MCD will be used to illustrate ridge regression, and the ALS data set will be used in Section 5.5.3.

Variables used in the analysis are as follows:

(Y): LEAFWT = total weight (grams) of photosynthetic material as derived from the actual harvesting of mesquite

(X1): DIAM1 = canopy diameter (meters) measured along the longer axis of the bush

(X2): DIAM2 = canopy diameter (meters) measured along the shorter axis of the bush

(X3): TOTHT = total height (meters) of the mesquite bush

(X4): CANHT = canopy height (meters) of the mesquite bush

(X5): DENS = plant unit density (number of primary stems/plant unit)

It is desired to estimate LEAFWT using the other measurements. It is natural to consider a multiplicative model, since volume should be a weighted product of these variables; hence, logarithms were taken of all variables and a log-linear model used throughout (see Chapter 6). It is natural to expect multicollinearity among the variables, since all but one are measures of size; hence, variable selection and/or biased estimation procedures are indicated.

The first step is to perform the usual least-squares regression involving all variables; results are presented in the top left portion of Table 5.13. As expected, there is a highly significant overall regression, while only a small number of the individual

TABLE 5.12 Mesquite Data for Section 5.5.2

Obs.	Group	DIAM1	DIAM2	TOTHT	CANHT	DENS	LEAFWT
1	MCD	1.80	1.15	1.30	1.00	1	401.3
2	MCD	1.70	1.35	1.35	1.33	1	513.7
3	MCD	2.80	2.55	2.16	0.60	1	1179.2
4	MCD	1.30	0.85	1.80	1.20	1	308.0
5	MCD	3.30	1.90	1.55	1.05	1	855.2
6	MCD	1.40	1.40	1.20	1.00	1	268.7
7	MCD	1.50	0.50	1.00	0.90	1	155.5
8	MCD	3.90	2.30	1.70	1.30	2	1253.2
9	MCD	1.80	1.35	0.80	0.60	1	328.0
10	MCD	2.10	1.60	1.20	0.80	1	614.6
11	MCD	0.80	0.63	0.90	0.60	1	60.2
12	MCD	1.30	0.95	1.35	0.95	1	269.6
13	MCD	1.20	0.90	1.40	1.20	1	448.4
14	MCD	1.50	0.70	1.00	0.70	1	120.4
15	MCD	2.80	1.70	1.70	1.20	1	378.7
16	MCD	1.40	0.85	1.50	1.10	1	266.4
17	MCD	1.50	0.60	0.65	0.64	1	138.9
18	MCD	2.40	2.40	1.50	1.20	1	1020.8
19	MCD	1.90	1.55	1.70	1.20	1	635.7
20	MCD	2.30	1.60	1.70	1.30	1	621.8
21	MCD	2.10	1.70	1.50	1.00	1	579.8
22	MCD	2.40	1.30	1.50	0.90	2	326.8
23	MCD	1.00	0.40	1.20	1.00	1	66.7
24	MCD	1.30	0.60	0.70	0.50	1	68.0
25	MCD	1.10	0.70	1.20	0.90	1	153.1
26	MCD	1.30	1.20	0.80	0.60	1	256.4
27	ALS	2.50	2.30	1.70	1.40	5	723.0
28	ALS	5.20	4.00	3.00	2.50	9	4052.0
29	ALS	2.00	1.60	1.70	1.40	1	345.0
30	ALS	1.60	1.60	1.60	1.30	1	330.9
31	ALS	1.40	1.00	1.50	1.10	1	163.5

TABLE 5.12 (Continued)

Obs.	Group	DIAM1	DIAM2	TOTHT	CANHT	DENS	LEAFWT
32	ALS	3.20	1.90	1.90	1.50	3	1160.0
33	ALS	1.90	1.80	1.10	0.80	1	386.6
34	ALS	2.40	2.40	1.60	1.10	3	693.5
35	ALS	2.50	1.80	2.00	1.30	7	674.4
36	ALS	2.10	1.50	1.25	0.85	1	217.5
37	ALS	2.40	2.20	2.00	1.50	2	771.3
38	ALS	2.40	1.70	1.30	1.20	2	341.7
39	ALS	1.90	1.20	1.45	1.15	2	125.7
40	ALS	2.70	2.50	2.20	1.50	3	462.5
41	ALS	1.30	1.10	0.70	0.70	1	64.5
42	ALS	2.90	2.70	1.90	1.90	1	850.6
43	ALS	2.10	1.00	1.80	1.50	2	226.0
44	ALS	4.10	3.80	2.00	1.50	2	1745.1
45	ALS	2.80	2.50	2.20	1.50	1	908.0
46	ALS	1.27	1.00	0.92	0.62	1	213.5

inflation factors are of sufficient magnitude to indicate a rather high degree of multicollinearity.

All possible combinations of independent variables were considered in the variable selection, although in this case all step procedures provide identical results. A C_p plot (not reproduced here) suggests a two-variable model using DIAM2 and CANHT; the resulting regression is summarized in the bottom left portion of Table 5.13. As expected, the residual mean square is decreased somewhat and the remaining coefficients show a much higher degree of statistical significance.

A ridge regression was next performed using values of k from 0 to 1 in steps of 0.025. The resulting ridge plot is reproduced in Figure 5.3. These are plots of the *standardized* coefficients, and thus the coefficients given for k = 0 (the least squares coefficients)

TABLE 5.13 Summary of Regressions

	Data set					
	MCD			ALS		
	Full Model					
Analysis of variance:						
Source	d.f.	MS	F	d.f.	MS	F
Total	25	0.7398		19	0.9248	
Regression	5	3.349	38.22	5	3.110	21.55
Residual	20	0.0876		14	0.1443	
		R = 0.9053			R = 0.8850	
Coefficients:						
	$\hat{\beta}$	t	VIF	$\hat{\beta}$	t	VIF
Intercept	5.409	--	--	4.344	--	--
DIAM1	0.4253	1.39	3.88	0.8601	1.19	8.77
DIAM2	1.135	4.97	3.81	1.057	2.26	4.95
TOTHT	0.2376	0.72	2.96	1.110	1.45	8.50
CANHT	0.5381	1.83	2.10	-0.5626	-0.77	7.83
DENS	-0.2855	-0.79	1.32	0.0563	0.33	1.84
	Selected Variables Model					
Analysis of variance:						
Source	d.f.	MS	F	d.f.	MS	F
Total	25	0.7398		19	0.9248	
Regression	2	8.261	96.31	2	7.596	54.29
Residual	23	0.0858		17	0.1399	
Coefficients:						
	$\hat{\beta}$	t	VIF	$\hat{\beta}$	t	VIF
Intercept	5.660	--	--	4.740	--	--
DIAM2	1.430	11.69	1.33	1.503	5.13	3.82
TOTHT	--	--	--	1.018	2.78	3.82
CANHT	0.7021	3.32	1.33	--	--	--

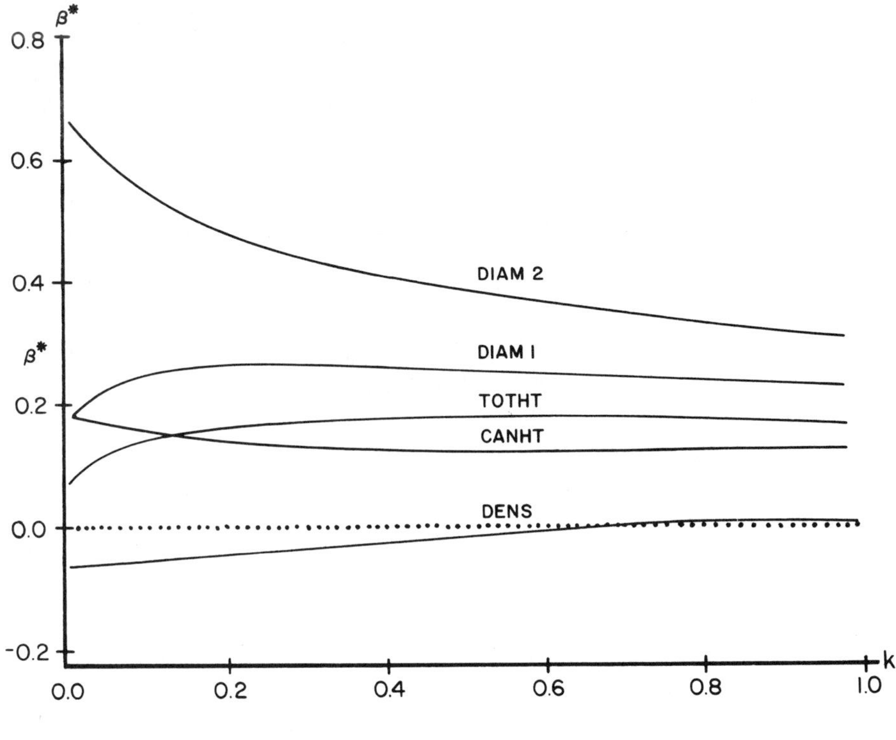

Figure 5.3 Ridge plot.

are not numerically the same as the coefficients given in Table 5.13. The ridge plot exhibits features usually found in plots of this type in that coefficients eventually tend to zero as k becomes larger. Some coefficients, in this case those for DIAM2 and TOTHT, initially increase before tending to zero. The largest single coefficient, that for DIAM2, decreases quite rapidly, while the coefficient for DENS is unimportant. As indicated earlier, it is difficult to define the k-value where the coefficients settle down. All coefficients, except that for DIAM2, appear to settle down in the neighborhood of k = 0.3, while that coefficient never seems to settle down. For purposes of this example, k = 0.4 was arbitrarily chosen.

The resulting coefficients, converted to the original scale of measurements, are given in the third line of the top half of Table

TABLE 5.14 Estimated Equations Coefficients/(F-values)[a]

Eq.	Intercept	DIAM1	DIAM2	TOTHT	CANHT	DENS
			MCD Data			
All vars.	5.409	0.4253	1.135	0.2376	0.5308	-0.2855
		(1.93)	(24.69)	(0.52)	(3.36)	(0.62)
Selected	5.660	--	1.430	--	0.7022	--
			(136.7)		(11.01)	
Ridge	5.297	0.5806	0.6908	0.4708	0.3763	-0.1196
		(4.58)	(11.71)	(2.10)	(1.42)	(0.07)
			ALS Data			
All vars.	4.344	0.8601	1.057	1.110	-0.5626	0.0562
		(1.41)	(5.11)	(2.10)	(0.59)	(0.11)
Selected	4.740	--	1.503	1.018	--	--
			(26.37)	(7.74)		
Ridge	4.742	0.6831	0.7451	0.5116	0.2401	0.1192
		(2.52)	(5.04)	(1.28)	(0.34)	(0.50)

[a]F-values for ridge regression computed as if ordinary least squares had been used.

5.14. Comparison of these with the coefficients of the least-squares equation (first line) shows that the coefficients are now of more nearly equal magnitude, that is, the variables are contributing more equally to the estimation of biomass. The F-values under the coefficients are computed in the same manner as for ordinary least squares; they provide only an indicator of the *relative* contribution of each coefficient but should not be used as indicators of statistical significance. The variables are contributing somewhat more equally to predictive ability, but there is a reversal in that *total* height appears more important than *canopy* height. Finally, the R^2 of the ridge regression is 0.762 compared to an R^2 of 0.905 for the least-squares equation and 0.885 for the two-variable selected equation. Thus the ridge regression has "lost" considerably with respect to the estimation of biomass for the data. It should be noted that the use of k = 0.3 would have provided a somewhat higher R^2 with relatively similar coefficients. Thus, possibly the choice of k = 0.4 is not optimum.

5.5.3 Use of Ridge Regression for Prediction

Prediction is defined to be the use of an estimated regression equation to predict values of the dependent variable for observations *not* in the data set used for estimation of the coefficients. One of the major reasons given for using variable selection or biased estimation procedures is that the resulting equations may provide better means for *prediction* as opposed to *estimation*, since the existence of multicollinearity tends to inflate the variances of predicted values [Walls and Weeks (1969)]. The data of this example provide an excellent illustration of this type of situation, since the two sets ALS and MCD describe the same physical phenomenon, but under not completely identical conditions. This section compares the results obtained using, respectively, all variables, selected variables, and ridge regression in each of the data sets.

The regression analyses using all variables for both data sets are given in Table 5.13. The results of the two analyses are similar; however, there are some differences. For the ALS data set, the TOTHT appears to be more important than the CANHT; in fact, the coefficient of the latter is negative, though not statistically significant. This feature of the ALS data is further amplified by the results of variable selection (bottom, Table 5.13) which retains the DIAM2 and TOTHT; the variance inflation factors are somewhat larger than those of the MCD data.

The results of the ridge regression analysis are given in Table 5.13. Ridge regression has performed moderately well in the sense that the equations for the two data sets are more similar than they are for either of the other two estimation methods (Table 5.13). Furthermore, the effects of all size measurements are more *diffused*, that is, they do not concentrate on only one or two factors.

Finally, of the height variables, total height is now more important for both data sets.

The final test of the effectiveness of the estimation methods is the use of the equations for prediction. In Table 5.15 are given summaries of the magnitudes of residuals when the equations from data set MCD are used to predict biomass for the ALS set, and vice versa. Statistics on both means, the variances, and the mean square errors are given, since the prediction error consists of both variation and bias.

It appears that ridge regression has improved prediction somewhat over the results obtained either by using all variables or the selected variables. The error associated with prediction is, of course, still considerably larger than the error from the estimation, but it is a considerable improvement over using no equation at all. It is somewhat surprising that the selected-variables equations do not perform as well as the all-variables equations.

TABLE 5.15 Summary of Predictions[a]

Predict from	Estimation method	Predict to MCD (n = 26)			Predict to ALS (n = 20)		
		s^2	$\bar{e}$	MSE	s^2	$\bar{e}$	MSE
MCD	LS-all	0.0701	0	0.0701	0.2152	-0.3713	0.3531
	LS-select	0.0789	0	0.0789	0.1612	-0.5503	0.4640
	Ridge	0.0944	0	0.0944	0.1907	-0.2996	0.2805
ALS	LS-all	0.1635	0.5190	0.4329	0.1064	0	0.1064
	LS-select	0.1229	0.6098	0.4948	0.1252	0	0.1252
	Ridge	0.0909	0.4597	0.3022	0.1319	0	0.1319

[a] s^2 is not the same as given by regression ANOVA as (n - 1) d.f. was used here for *all* models.

5.6 SIGNIFICANCE TESTS

It has already been noted that the implications of statistical significance tests are changed when variable selection procedures are used. Although this fact is inherent in theory, it may not be readily appreciated by many users of regression methodology. This phenomenon is illustrated with a sampling experiment.

A sample consisted of 100 observations of a normally distributed dependent variable, y, with zero mean and unit variance and 25 associated rectangularly distributed variables x_1, x_2, ..., x_{25}, each with range 0 to 1. *All* variables were uncorrelated with each other; hence, all results which test "significant" will, in fact, be type I errors. One hundred such samples were generated.

Variable selections were performed using backward elimination on the 100 samples, using the 0.10 significance level for stopping the deletion procedure, producing equations with all remaining coefficients "statistically significant" at the 0.10 level.

Table 5.16 summarizes the results of the selections. It is seen that a (experiment-wise) type I error (*including* one or more coefficients or variables when none should be) was committed more than 90% of the time. In fact, approximately 60% of the resulting equations included *three* or *more* "significant" coefficients.

The number of individual coefficients included in an equation could be considered as a binomial trial with the selection of a variable considered a "success"; the probability of a success is 0.10 and the number of trials is 25. However, the resulting distribution given as "Expected no." in Table 5.16 does not fit the actual frequencies of selected variables very well, as indicated by a χ^2 goodness-of-fit test which gives a highly significant χ^2 value of 67.66 with 9 degrees of freedom.

The point of this example is that the type I error protection against erroneous rejection is not even truly based on the *total* number of variables in the original model. This result implies that procedures which start selection with "everything but the kitchen sink" will have greater chances of erroneously including variables than those which start with a carefully chosen set. The latter, of course, may have correspondingly greater chances of *not* including important variables.

TABLE 5.16 Number of Variables Selected by Two Methods

No. var. selected	No. of equations	Expected no.
0	9	7.2
1	18	20.0
2	14	26.6
3	16	22.6
4	16	13.8
5	10	6.5
6	6	2.4
7	8	0.7
8	2	0.2
9	1	0.1
Average	3.25	2.50

5.7 PROBLEMS

1. Given the following (corrected) $X'X$, $X'\underline{y}$, $(X'X)^{-1}$, and $\underline{y}'\underline{y}$ matrices, assuming 20 observations:

$$X'X = \begin{bmatrix} 2 & -1 & 0 \\ -1 & 4 & 3 \\ 0 & 3 & 3 \end{bmatrix} \qquad X'\underline{y} = \begin{bmatrix} 6 \\ 7 \\ 9 \end{bmatrix}$$

$$(X'X)^{-1} = \begin{bmatrix} 1 & 1 & -1 \\ 1 & 2 & -2 \\ -1 & -2 & 7/3 \end{bmatrix} \qquad \underline{y}'\underline{y} = [60]$$

(a) Obtain $\hat{\underline{\beta}}$, the analysis of variance for H_0: $\underline{\beta} = 0$, and the F for each H_{0i}: $\beta_i = 0$.

(b) Perform a step-up selection to include all variables.

(c) Perform a step-down selection to the one-variable equation.

(d) Comment on results.

2. Regression analyses have been calculated for all possible variable combinations of a five-variable data set. Use the data given in Table 5.17 to simulate a step-up and a step-down selection procedure (but do not do significance tests). Prepare a C_p plot; use formula (5.5).

3. Surveys of low-income Mexican Americans indicate that members of this group see education as the vehicle by which they or their children can escape poverty. However, this group continues to have a very high dropout rate from both high school and college. The purpose of this study is to test the effect of selected variables on the high percentage of Mexican Americans who fail to continue their formal education beyond age 16.

   The data for 53 census tracts were taken or derived from the U.S. Bureau of Census:

   Census of Population and Housing, 1970

   Census Tracts El Paso, Texas, SMSA

   The variables are

   DROPS = % of population age 16-21 not in school (dependent variable)
   INC = median family income
   HOUS = % of housing units with more than 1 person per room
   GRADS = % of population over age 25 who are high school graduates
   OWN = % of housing units which are owner occupied
   SPAN = % of population classed as of Spanish surname
   UNEMP = civilian unemployment rate
   LT16 = % of population under age 16

TABLE 5.17 Variables Included[a]

1	2	3	4	5	SSE
x	x	x	x	x	130
x	x	x	x		131
	x	x	x	x	133
x	x		x	x	152
x	x	x		x	1302
x		x	x	x	3089
	x		x	x	152
x	x		x		162
	x	x	x		163
	x	x		x	2127
x	x			x	2252
x	x	x			3650
x			x	x	3928
x		x	x		13910
x		x		x	14010
		x	x	x	134229
	x		x		178
	x			x	3300
	x	x			3658
x	x				10592
x			x		15119
x		x			17335
x				x	18174
			x	x	147183
		x		x	176208
		x	x		600131
x					18438
	x				18673
		x			646766
			x		827825
				x	843767

[a]SS Total = 1,000,000; n = 13.

Table 5.18 gives information on a number of the best equations as obtained from examining all possible regressions.

TABLE 5.18 Data for Variable Selection

Var.	Rank	INC	GRAD	SPAN	HOUS	UNEM	OWN	LT16	R^2	C_p
1	Best	x							0.362	2.20
	2		x						0.318	5.80
	3						x		0.226	13.33
2	Best		x	x					0.401	1.01
	2	x				x			0.380	2.73
	3	x		x					0.372	3.38
	4	x	x						0.366	3.87
	5	x						x	0.364	4.04
	6	x					x		0.363	4.12
	7	x			x				0.363	4.12
3	Best	x	x	x					0.422	1.29
	2		x	x		x			0.416	1.78
	3		x	x	x				0.407	2.52
	4		x	x				x	0.406	2.60
	5		x	x			x		0.406	2.60
	6	x		x		x			0.394	3.58
	7	x	x			x			0.383	4.48
4	Best	x	x	x		x			0.446	1.36
	2	x	x	x				x	0.427	2.89
	3	x	x	x	x				0.424	3.14
	4		x	x	x	x			0.423	3.22
	5	x	x	x			x		0.422	3.24
	6		x	x		x	x		0.420	3.45
	7		x	x		x		x	0.419	3.50
5	Best	x	x	x		x		x	0.447	3.21
6	Best	x	x	x		x	x	x	0.449	5.05
7	--	x	x	x	x	x	x	x	0.450	7.00

(a) Simulate a backward elimination stopping when all variables are significant at 0.10 level. (Use R^2 values. Remember that all tests are invariant through scale transformations; thus, R^2 = SSR if SST is assumed to be 1.0.)
(b) Simulate forward selection, stopping when next added variable is not significant at 0.10 level.
(c) Make a C_p plot and recommend a most suitable equation. Qualify choice if desirable.

4. "Road Test" articles in Motor Trend magazine for March-July, 1974, provide data on factors affecting gas mileage of various 1974 cars. Five variables are used to estimate gas mileage (MILES):

ESIZE: engine size (cubic inches)

HP: engine horse power

BARR: number of carburetor barrels

WT: weight of car (pounds)

TIME: time (seconds) for quarter mile (inverse of speed)

The data are given in Table 5.19. A linear regression analysis is performed. Selected results (all SS and SP corrected for mean) are given in Table 5.20.
(a) Complete an analysis of this set of data. Perform all relevant tests and interpretations. Note that the residuals from this analysis are also given in Table 5.20.
(b) All possible regressions were performed using subsets of these five variables. The results are as follows:

Var.	R^2	C_p
TIME	0.1753	101.788
BARR	0.3035	81.457
HP	0.6024	34.050
ESIZE	0.7183	15.668
WT	0.7528	10.203

Var.	R^2	C_p
BARR TIME	0.3093	82.553
HP BARR	0.6046	35.716
HP TIME	0.6369	30.586
ESIZE TIME	0.7216	17.157
ESIZE HP	0.7482	12.626
ESIZE BARR	0.7737	8.889
ESIZE WT	0.7809	7.742
BARR WT	0.7924	5.917
WT TIME	0.8264	0.531
HP WT	0.8268	0.468
HP BARR TIME	0.6369	32.586
ESIZE HP TIME	0.7542	13.981
ESIZE HP BARR	0.2740	10.846
ESIZE BARR TIME	0.7837	9.298
ESIZE BARR WT	0.8182	3.835
ESIZE WT TIME	0.8264	3.529
ESIZE HP WT	0.8268	2.461
HP BARR WT	0.8271	2.428
BARR WT TIME	0.8272	2.404
HP WT TIME	0.8348	1.203
ESIZE HP BARR TIME	0.7844	11.185
ESIZE HP BARR WT	0.8276	4.345
ESIZE BARR WT TIME	0.8278	4.309
HP BARR WT TIME	0.8348	3.200
ESIZE HP WT TIME	0.8351	3.144
ALL VAR --	0.8361	5.000

What subset(s) would you choose? Explain your reasoning.

(c) Complete regressions have been performed on five selected subsets; the results are summarized on page 158. Do these results confirm or change your earlier decision?

TABLE 5.19 Car Gas-Mileage Data

Obs.	Name	ESIZE	HP	BARR	WT	TIME	MILES	RESIDUAL
1	Mazda Rx-4	160.0	110	4	2620	16.46	21.0	-1.85
2	Mazda Rx-4 Wagon	160.0	110	4	2875	17.02	21.0	-0.93
3	Datsun 710	108.0	93	1	2320	18.61	22.8	-2.21
4	Hornet 4 Drive	258.0	110	1	3215	19.44	21.4	-0.35
5	Hornet Sport	360.0	175	2	3440	17.02	18.7	0.21
6	Valiant	225.0	105	1	3460	20.22	18.1	-2.80
7	Duster 360	360.0	245	4	3570	15.84	14.3	-1.62
8	Mercedes 240D	146.7	62	2	3190	20.00	24.4	1.57
9	Mercedes 230	140.8	95	2	3150	22.90	22.8	-1.34
10	Mercedes 280	167.6	123	4	3440	18.30	19.2	-0.47
11	Mercedes 280C	167.6	123	4	3440	18.90	17.8	-2.27
12	Mercedes 450SE	275.8	180	3	4070	17.40	16.4	1.34
13	Mercedes 450SL	275.8	180	3	3730	17.60	17.3	0.38
14	Mercedes 450SLC	275.8	180	3	3780	18.00	15.2	-1.74
15	Cadillac	472.0	205	4	5250	17.98	10.4	-0.22
16	Lincoln	460.0	215	4	5424	17.82	10.4	1.10
17	Imperial	440.0	230	4	5345	17.42	14.7	5.77
18	Fiat 128	78.7	66	1	2200	19.47	32.4	5.78
19	Honda Civic	75.7	52	2	1615	18.52	30.4	0.87
20	Toyota Corolla	71.1	65	1	1835	19.90	33.9	5.18

21	Toyota Corona	120.1	97	1	2465	20.01	21.5	-3.71
22	Challenger	318.0	150	2	3520	16.87	15.5	-2.76
23	Javelin	304.0	150	2	3435	17.30	15.2	-3.67
24	Camaro	350.0	245	4	3840	15.41	13.3	-0.89
25	Firebird	400.0	275	2	3845	17.05	19.2	2.45
26	Fiat X1-9	79.0	66	1	1935	18.90	27.3	-0.28
27	Porsche 914-2	120.3	91	2	2140	16.70	26.0	0.94
28	Lotus Europa	95.1	113	2	1513	16.90	30.4	2.74
29	Pantera	351.0	264	4	3170	14.50	15.8	-0.73
30	Ferrari Dino 197	145.0	175	6	2770	15.50	19.7	-0.64
31	Maserati Bora	301.0	335	8	3570	14.60	15.0	1.40
32	Volvo 142E	121.0	109	2	2780	18.60	21.4	-1.26

TABLE 5.20 Selected Computations

Var.	Mean	Corrected SS	Elements of $X'\underline{y}$	Diagonal of $(X'X)^{-1}$	Elements of $(X'X)^{-1}X'\underline{y}$
X_1: ESIZE	2.3072×10^2	4.7619×10^5	-1.9626×10^4	3.9751×10^{-5}	7.5181×10^{-3}
X_2: HP	1.4669×10^2	1.4573×10^5	-9.9427×10^3	6.0618×10^{-5}	-2.3743×10^{-2}
X_3: BARR	2.1825	8.0875×10	-1.6626×10^2	7.2155×10^{-2}	2.7186×10^{-1}
X_4: WT	3.2173×10^3	2.9679×10^7	-1.5862×10^5	4.4272×10^{-7}	-5.0725×10^{-3}
X_5: TIME	1.7849×10	9.8988×10	1.3978×10^2	4.6885×10^{-2}	6.6907×10^{-1}
y: MILES	2.0452×10	1.1261×10^3	--	--	--

Regr.	Var.	Diag. C	$(X'X)^{-1}X'\underline{y}$
1	WT	3.36941×10^{-8}	-5.34447×10^{-3}
2	HP	1.21228×10^{-5}	-3.17730×10^{-2}
	WT	5.95249×10^{-8}	-3.87783×10^{-3}
3	WT	3.47551×10^{-8}	-5.04798×10^{-3}
	TIME	1.04203×10^{-2}	9.29198×10^{-1}
4	BARR	1.51315×10^{-2}	-8.21523×10^{-1}
	WT	4.12336×10^{-8}	-4.46458×10^{-3}
5	HP	3.37752×10^{-5}	-1.78223×10^{-2}
	WT	8.52611×10^{-8}	-4.3588×10^{-3}
	TIME	2.90318×10^{-2}	5.10834×10^{-1}

Chapter 6

MODELS NOT STRICTLY LINEAR

6.1 INTRODUCTION

In Chapter 3 it was indicated that the word *linear* in multiple linear regression models refers to the fact that the model is linear in the parameters. It was also shown that linear models can be used to estimate relationships which are not necessarily linear in the variables, and some examples of such regressions have been presented. This chapter provides additional methodologies for use in the estimation of relationships which are not strictly linear. The reader is advised at this stage that a review of the analysis of factorial experiments is useful for a better understanding of Section 6.5.

One of the most frequently used representations of nonlinear relationships is afforded by a polynomial equation; this is discussed in Section 6.2. The extreme popularity of polynomial equations has given rise to a specialized technique using *orthogonal polynomials*, which is presented in Section 6.3. Extension of polynomial equations to two or more variables is discussed in Sections 6.4 and 6.5.

Another popular device used to obtain estimates for nonlinear relationships is to perform transformations on the dependent variable. This is discussed in Section 6.6. Models which cannot be linearized by use of polynomials or transformations require the application of nonlinear least squares methodology which is beyond the scope of this book, but it is briefly discussed in Section 6.6.

6.2 POLYNOMIAL REGRESSION

The polynomial regression model with one independent variable is

$$y = \beta_0 + \beta_1 x + \beta_2 x^2 + \cdots + \beta_m x^m + \varepsilon$$

This particular type of relationship is attractive to users because it is easy to implement, and also because it is possible to approximate almost any function, whose exact form may be unknown, by a Taylor's series expansion, which is a form of polynomial function. Also a polynomial function is quite attractive, since it is simply a sum of linear, quadratic, cubic, or higher terms, whose corresponding shapes are well known and illustrated in Figure 6.1. A knowledge of these shapes can be quite useful in interpreting the results of response surface experiments such as are discussed in Section 6.5.

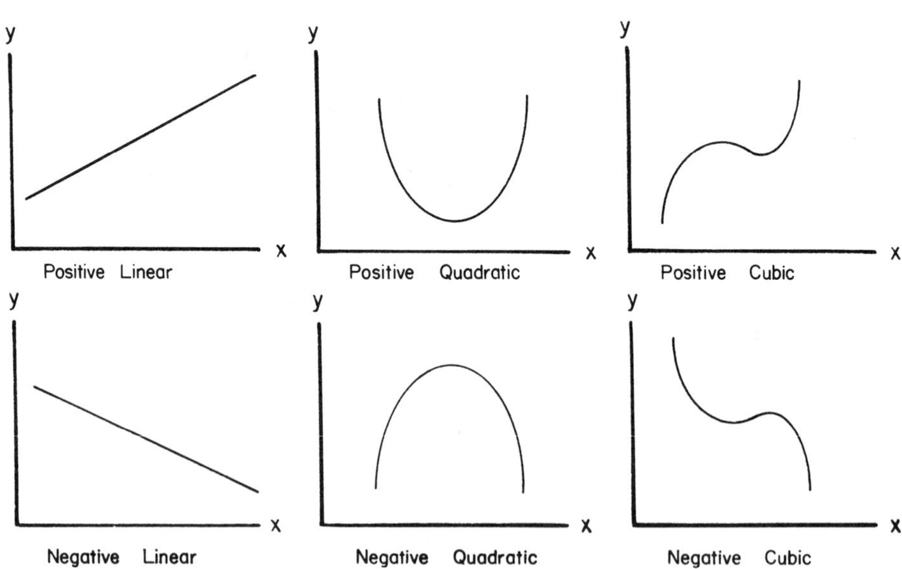

Figure 6.1 Typical polynomial curves.

As discussed in Chapter 2, a redefinition of variables makes a polynomial amenable to the usual linear regression procedures (estimates and tests) without modification. Some special problems and restrictions should, however, be kept in mind when using polynomial regression:

1. The numbers involved in computations may become quite large (or quite small), since they will be the sums of higher powers of observed values. In fact, the numbers may become unmanageable even for computers with floating decimal capabilities and may provide numerically inaccurate results. A transformation of the scale of x, preferably such that $0.1 < x < 1.0$, is advisable.
2. In many cases, the polynomial model is used as an approximation to some other functional form. This approximation is valid *only* in the region of the data; hence, extrapolation is not advisable.

The danger of extrapolation is illustrated by the following example. Data points were generated according to the model

$$y = x^{0.5} + \varepsilon$$

for values of x from 2 to 10, with ε being a normally distributed random variable with mean zero and standard deviation 0.5 (Table 6.1). Quadratic polynomial regressions were performed using the first seven, and then all of the observations, resulting, respectively, in the following estimated equations and estimated standard deviations:

$$\hat{\mu}_1 = 0.109 + 0.668x - 0.0500x^2 \qquad s_1 = 0.446$$

$$\hat{\mu}_2 = 0.424 + 0.436x - 0.0192x^2 \qquad s_2 = 0.435$$

The resulting values are given in Table 6.1 and compared with the population values ($\mu = x^{0.5}$). For the first six observations, both equations appear to provide comparable and satisfactory estimates, after which they diverge, with the extrapolated values ($\hat{\mu}_1$) becoming much poorer estimates of the true values. In fact, the extrapolated

TABLE 6.1 Example of Polynomial Approximation

x	y	μ	$\hat{\mu}_1$	$\hat{\mu}_1 - \mu$	$\hat{\mu}_2$	$\hat{\mu}_2 - \mu$
1	1.152	1.000	0.726	-0.274	0.841	-0.159
2	0.605	1.414	1.242	-0.172	1.219	-0.195
3	1.580	1.732	1.656	-0.076	1.559	-0.173
4	2.039	2.000	1.968	-0.032	1.861	-0.139
5	2.408	2.236	2.179	-0.057	2.124	-0.112
6	2.554	2.449	2.228	-0.221	2.349	-0.100
7	2.018	2.646	2.295	-0.351	2.536	-0.110
8	3.031	2.828	2.201[a]	-0.627	2.685	-0.143
9	2.310	3.000	2.005[a]	-0.995	2.795	-0.205
10	3.138	3.162	1.707[a]	-1.455	2.866	-0.296

[a]Extrapolated values.

curve starts decreasing after x = 7, a phenomenon which cannot occur with the correct (exponential) function.

6.3 ORTHOGONAL POLYNOMIALS

For a large class of practical situations involving polynomial regression estimation, the use of *orthogonal polynomials* provides a simplified procedure for estimation and testing purposes. This procedure is generally applicable only if the following conditions apply:

1. The x variable occurs at equally spaced intervals
2. The minimum degree of the polynomial required to adequately describe the data is to be determined by a step-up procedure, that is, adding higher degree terms one at a time until further additions are deemed nonsignificant

The principle of the orthogonal polynomial procedure can be explained by considering it as a step-up procedure implemented by the use of residuals (Section 2.5). The first step is to estimate and test for the existence of a simple linear regression (using residuals from the means). In the second step, the coefficient and associated statistics for the quadratic term can be obtained by regressing the dependent variable on the residuals from the regression of x^2 on x. Continuing, the cubic term involves the regression of the dependent variable on the residuals from the regression of x^3 on x^2 and x, etc. These residuals are called *orthogonal polynomials* and are calculated for each successive term of the polynomial. The multiple regression of y on these residuals has the following properties:

1. The residuals corresponding to the individual polynomial terms are uncorrelated; hence, the regression for each term may be independently performed as a one-variable linear regression
2. Each of these individual regressions produces the coefficient and associated sum of squares for that term, adjusted for all terms involving *lower* powers of x. Hence, each of these regressions provides the statistics for the step-up significance test

When the observations on the independent variable are equally spaced, residuals can be calculated once and for all for any given sample size, regardless of the actual values of x, since the scale and origin of variables have no effect on the statistics generated by a linear regression. Such residuals have indeed been calculated and tabled, together with procedures to reconvert analyses to any original x-scale. An extensive table, including procedures for derivation and implementation, is given in Anderson and Housemann (1942). Smaller tables are included in many textbooks and books of statistical tables; such a table is reproduced in the appendix. In most references, the orthogonal polynomials are denoted by the Greek letter ξ; in an effort to maintain the custom of reserving Greek letters for parameters, this book will refer to them as z-values.

In order to facilitate the use of orthogonal polynomials, most tables have the following features:

1. Polynomial values are multiplied by a scale factor such that all values are integers
2. Since they are residuals, all polynomial values have zero mean; hence, correction for the mean is not needed, and, in fact, the intercept is the sample mean
3. The sum of squares of each of the polynomial values is provided

It should be noted that the use of the orthogonal polynomials to obtain a step-up regression analysis for a polynomial relationship does not provide *all* the statistics that would be provided by standard multiple regression methodology using the powers of x. The use of orthogonal polynomials *does* provide the following:

1. An estimating equation of the form

   $$\hat{\mu}_z = \hat{\alpha}_0 + \hat{\alpha}_1 z_1 + \hat{\alpha}_2 z_2 + \cdots + \hat{\alpha}_m z_m$$

   where the z_i are the orthogonal polynomial values corresponding to x^i, which produces the same $\hat{\mu}_x$-values as the m-th order polynomial using the original variables, and
2. An easily implemented step-up procedure for estimating the effectiveness of the successive degrees of polynomial regression estimation, by providing the sums of squares for significance tests

However, the use of the orthogonal polynomials does *not* provide the following:

1. The actual coefficients of the polynomials in the originally observed variables. These may be obtained by use of formulas supplied, for example, in Anderson and Housemann (1942), but the associated computations are tedious
2. Statistical significance tests for partial coefficients of the polynomial equation in the original variables, since the only statistical tests afforded are those for the overall regression and for the last (highest order) term added in the step-up procedure

3. The estimation of equations which *exclude* lower order polynomial terms. Thus, for example, it is not possible to use the tabled orthogonal polynomials to estimate an equation of the form

$$y = \beta_0 + \beta_1 x + \beta_2 x^3 + \beta_3 x^4 + \varepsilon$$

The regression using only the orthogonal polynomial values corresponding to the linear, cubic, and fourth powers does *not* correspond to the equivalent polynomial in the original variable

4. Estimates of the dependent variable $\hat{\mu}_x$ for values of the independent variable not originally observed. The orthogonal polynomial method cannot readily be used for interpolation or extrapolation

The orthogonal polynomial tables may also be used in situations where there are multiple observations at each of the equally spaced x-values; however, the polynomials remain uncorrelated only if there are equal sample sizes at each x-value. In such cases the mean (or totals) may be used as the observations; the coefficients and resulting sums of squares must be properly adjusted for sample size: sums of squares must be divided or multiplied by n (number of replications) when totals or means, respectively, are used, and coefficients divided by n if totals are used.

Orthogonal polynomials may also be used if the data have equally spaced values of some function of x, such as x^i or log i, although in this case the true polynomial will be in terms of the function of x rather than of x itself.

As an example, consider estimated monthly retail sales in the United States for 1964. The data are given in Table 6.2. It is desired to obtain a polynomial curve to fit these data; the model is

$$y = \beta_0 + \beta_1 x + \beta_2 x^2 + \beta_3 x^3 + \beta_4 x^4 + \beta_5 x^5 + \varepsilon$$

where

TABLE 6.2 Estimated Monthly Sales of Retail Stores, 1964

Month	Sales ($\$ \times 10^9$)	Orthogonal polynomials (z-values)				
		Linear	Quadratic	Cubic	Fourth degree	Fifth degree
Jan.	19.15	-11	55	-33	33	-33
Feb.	18.76	-9	25	3	-27	57
Mar.	20.50	-7	1	21	-33	21
Apr.	21.19	-5	-17	25	-13	-29
May	22.51	-3	-29	19	12	-44
June	22.24	-1	-35	7	28	-20
July	22.15	1	-35	-7	28	20
Aug.	21.78	3	-29	-19	12	44
Sept.	21.31	5	-17	-25	-13	29
Oct.	22.61	7	1	-21	-33	-21
Nov.	21.70	9	25	-3	-27	-57
Dec.	27.72	11	55	33	33	33
Sum of squares		572	12,012	5148	8008	15,912

y = sales

x = month (January = 1, February = 2, ..., December = 12)

The orthogonal polynomials z_i corresponding to the first five powers of x (for 12 observations) are also given in Table 6.2. Note that the table in the appendix gives only the "bottom half" of the polynomial values; the "top half" is a *mirror* image with reversed signs for the odd-powered polynomials. If the sample size is odd, the center value is unique. The corresponding model for the orthogonal polynomial regression is

$$y = \mu + \alpha_1 z_1 + \alpha_2 z_2 + \alpha_3 z_3 + \alpha_4 z_4 + \alpha_5 z_5 + \varepsilon$$

The resulting matrices for solving the normal equations are

$$Z'Z = \begin{bmatrix} 572 & & & & \\ & 12012 & & & \\ & & 5148 & & \\ & & & 8008 & \\ & & & & 15192 \end{bmatrix}$$

$$Z'\underline{y} = \begin{bmatrix} 133.8197 \\ 71.89954 \\ 241.1798 \\ 253.5598 \\ 40.09770 \end{bmatrix}$$

Clearly Z'Z is a diagonal matrix with the elements Σz_i^2; hence, $(Z'Z)^{-1}$ is a diagonal matrix with elements $1/\Sigma z_i^2$, and each coefficient and corresponding sum of squares is obtained separately as in one-variable (total) regression. Thus,

$$\hat{\alpha}_1 = \frac{\sum z_1 y}{\sum z_1^2} = \frac{133.8197}{572} = 0.2340$$

and the corresponding sum of squares due to α_1 is

$$SS(\alpha_1) = \frac{(\sum z_1 y)^2}{\sum z_1^2} = 31.3071$$

Similarly,

$$\hat{\alpha}_2 = 0.005986 \qquad SS(\alpha_2) = 0.4304$$

$$\hat{\alpha}_3 = 0.04685 \qquad SS(\alpha_3) = 11.2991$$

$$\hat{\alpha}_4 = 0.03166 \qquad SS(\alpha_4) = 8.0285$$

$$\hat{\alpha}_5 = 0.002544 \qquad SS(\alpha_5) = 0.1010$$

and since the coefficients are independent,

$$SS(\underline{\alpha}) = \sum SS(\alpha_i) = 51.1661$$

As indicated before, the sums of squares due to the α_i are used for a step-up analysis; the F-statistics of the analysis of variance in Table 6.3 must be used sequentially. Thus, the first test is for the hypothesis

$$H_0: \alpha_1 = 0 \qquad \text{(no linear regression)}$$

which is rejected with F(1, 10) = 31.3071/2.3791 = 13.159. The second test is for

$$H_0: \alpha_2 = 0 \qquad \text{(no quadratic, adjusted for linear)}$$

which is not rejected with F(1, 9) = 0.166. It is normally recommended that the step-up procedure be stopped when *two successive* terms are nonsignificant; hence, we make the third test for

$$H_0: \alpha_3 = 0 \qquad \text{(no cubic, adjusted for linear and quadratic)}$$

which is rejected with F(1, 8) = 7.495. Likewise, the fourth degree term is included, while the fifth degree is not. Actually, a sixth degree could have been attempted, but tables for orthogonal polynomials for degrees higher than five are not generally available. Note that at each step the residual mean square for that step is used as the error. In situations where there are multiple observations for each x-value, a lack-of-fit test (Section 4.2) is more appropriate.

TABLE 6.3 Step-up Analysis of Variance

Source	d.f.	SS	MS	F
Total	11	55.0977		
Due to α_1, Linear	1	31.3071	31.3071	13.159[a]
Residual	10	23.7906	2.3791	
Add. due to α_2, Quadratic	1	0.4304	0.4304	0.166
Residual	9	23.3602	2.5956	
Add. due to α_3, Cubic	1	11.2991	11.2991	7.495[b]
Residual	8	12.0611	1.5076	
Add. due to α_4, Fourth degree	1	8.0285	8.025	13.936[a]
Residual	7	4.0326	0.5761	
Add. due to α_5, Fifth degree	1	0.1010	0.1010	0.154
Residual	6	3.9316	0.6553	

[a]Significant at 0.01 level.
[b]Significant at 0.05 level.

The analysis thus suggests that a fourth degree polynomial is sufficient. Remember that, although α_2 was found to be nonsignificant, the use of orthogonal polynomials will not allow the dropping of the α_2 term.

The resulting fourth degree equation

$$\hat{\mu}_z = 21.802 + 0.2340z_1 + 0.005986z_2 + 0.04685z_3 + 0.03166z_4$$

provides the same estimated values ($\hat{\mu}_z$) as provided by the fourth degree polynomial model using the original x variable. The resulting curve can be graphed by substituting the z-values: for January (the first month)

$$\hat{\mu}_z = 21.802 + 0.2340(-11) + 0.005986(55) + 0.04685(-33) + 0.03166(33) = 19.056$$

compared to actual sales of 19.15. The fit of the equation is

TABLE 6.4 Actual and Estimated Sales

Month	Actual	Estimated
January	19.15	19.05
February	18.76	19.13
March	20.50	20.11
April	21.19	21.29
May	22.51	22.20
June	22.24	22.57
July	22.15	22.38
August	21.78	21.82
September	21.31	21.29
October	22.61	21.42
November	21.70	23.06
December	27.72	27.30

illustrated in Table 6.4 and Figure 6.2, which show the actual and estimated values. The fourth degree curve fits the data quite well except that the polynomial cannot adequately represent the spurt in sales in October and the lull in November.

For comparison purposes, the data were subjected to a fourth degree polynomial regression using the model

$$y = \beta_0 + \beta_1 x + \beta_2 x^2 + \beta_3 x^3 + \beta_4 x^4 + \varepsilon$$

with results given in Table 6.5. Superficially, the results bear little resemblance to the orthogonal polynomial regression. However, the sum of squares due to β_4 is essentially the same as for α_4, and the residual mean square is essentially equal to the residual mean square after estimating α_4. The minor differences observed in these results are due to rounding error, which often occurs in polynomial regressions. The estimated values are the same for both methods, that is, $\hat{\mu}_z = \hat{\mu}_x$; they are not reproduced here.

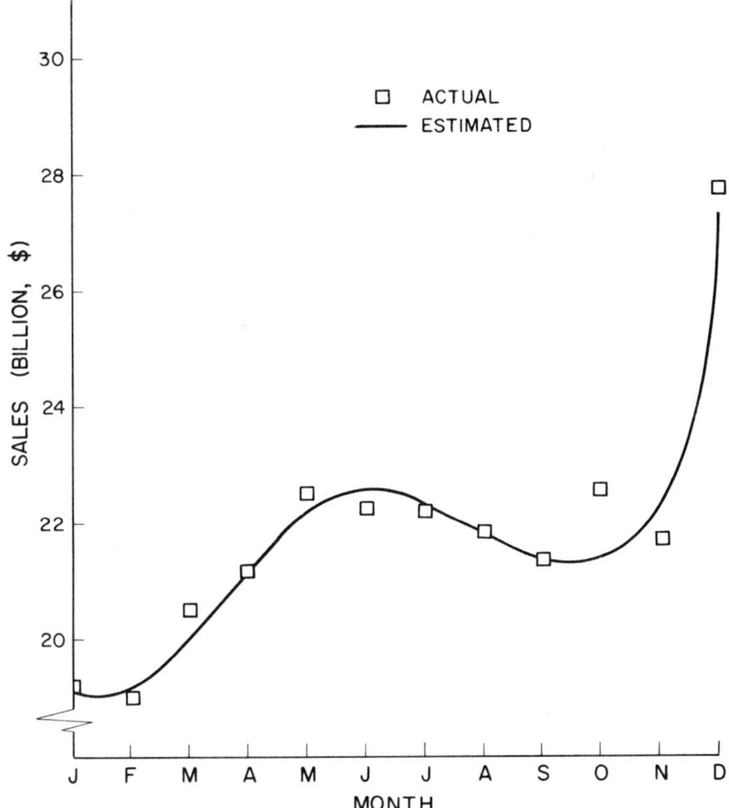

Figure 6.2 Actual and estimated sales.

TABLE 6.5 Polynomial Equation

Coefficient	Estimate	SS(adj)
$\hat{\beta}_0$	20.797	
$\hat{\beta}_1$	-3.0140	1.4123
$\hat{\beta}_2$	1.4711	3.8391
$\hat{\beta}_3$	-0.20858	5.9962
$\hat{\beta}_4$	0.0092237	8.0217
Residual mean square		0.57725

6.4 MULTIDIMENSIONAL POLYNOMIALS AND RESPONSE SURFACES

The techniques of polynomial regression extend to polynomial regression in several variables. Basic principles and computations of multiple polynomial regression are straightforward extensions of one-variable polynomials, but the additional variables provide more flexibility in the construction of the model and, consequently, somewhat greater complexities in interpretation.

For example, consider the two-variable model

$$y = \beta_0 + \beta_1 x_1 + \beta_2 x_2 + \beta_3 x_1 x_2 + \varepsilon$$

and assume the estimated equation to be

$$\hat{\mu}_x = 4 + x_1 + x_2 - 0.5 x_1 x_2$$

where x-values range from 0 to 4. Interpretation of this equation is aided by a rearrangement of terms:

$$\hat{\mu}_x = 4 + (1 - 0.5x_2)x_1 + x_2$$

The rearrangement of terms shows clearly that the slope in the x_1 direction varies with different values of x_2. In this example, if $x_2 = 2$, there is no response to x_1; if x_2 is greater than 2, the response to x_1 has a negative slope; while if x_2 is less than 2, the response to x_1 is positive. This phenomenon is caused by the term $\beta_3 x_1 x_2$, which, in fact, represents an interaction.

Interaction is usually associated with a factorial experiment [see, e.g., Snedecor and Cochran (1967, Chapter 12)], where it is defined to be the failure of the effects of one treatment to be consistent over levels of another treatment. In the present example, the nature of the interaction is more specific: the linear response due to x_1 changes linearly with respect to values of x_2; hence, it is known as the *linear by linear* interaction. This interaction is symmetric with respect to x_1 and x_2, which is seen by the alternate arrangement:

$$\hat{\mu}_x = 4 + x_1 + (1 - 0.5x_1)x_2$$

The generalization of this type of model to higher powers simply produces more interaction terms. For example, assume a polynomial of the form

$$y = \beta_0 + \beta_1 x_1 + \beta_2 x_2 + \beta_3 x_1^2 + \beta_4 x_2^2 + \beta_5 x_1 x_2 + \beta_6 x_1^2 x_2 + \beta_7 x_1 x_2^2 + \varepsilon$$

The β_5 coefficient is the linear by linear interaction. The β_7 coefficient, the linear by quadratic interaction, indicates that the slope in the x_1 direction changes with x_2^2, i.e., it changes more rapidly with x_2 as x_2 gets large numerically. The β_6 term quadratic by linear, indicates that the quadratic term, the curvature of the response to x_1, changes with x_2 and may, in fact, reverse from concave to convex or vice versa (see Figure 6.1).

It is obvious that there is an infinity of possibilities and that the resulting notation can become exceedingly awkward. A useful notational device is to use multiple subscripts on coefficients to identify the corresponding exponents on the variables. Using this, a two-factor quadratic polynomial is represented as

$$y = \beta_{00} + \beta_{10} x_1 + \beta_{01} x_2 + \beta_{20} x_1^2 + \beta_{02} x_2^2 + \beta_{11} x_1 x_2 + \beta_{12} x_1 x_2^2 + \beta_{21} x_1^2 x_2 + \beta_{22} x_1^2 x_2^2 + \varepsilon$$

For a three-factor polynomial of degree m, the notation can be described by the formula

$$y = \sum_{i=0}^{m} \sum_{j=0}^{m} \sum_{k=0}^{m} \beta_{ijk} x_1^i x_2^j x_3^k + \varepsilon$$

A polynomial need not contain all terms as indicated above; in fact, an m-th order polynomial is often defined to include only terms such that the *sum* of exponents of all variables in any term is less than or equal to m.

A two-variable polynomial is illustrated using monthly retail sales data for 12 years (1953-1964) (Table 6.6). The polynomial is used to estimate the monthly trend within years, the long-term (annual) trend, and the changes in the monthly trend over years.

TABLE 6.6 Estimated Sales of All Retail Stores

	Year											
Month	1953	1954	1955	1956	1957	1958	1959	1960	1961	1962	1963	1964
January	13.05	12.34	13.15	13.73	14.74	15.29	16.23	16.31	15.80	16.94	18.26	19.15
February	12.33	12.06	12.64	13.55	14.06	13.78	14.96	15.83	15.07	15.98	17.09	18.76
March	13.96	13.54	14.57	15.72	15.79	15.55	17.19	17.42	17.93	18.97	19.65	20.50
April	14.17	14.32	15.49	14.89	16.44	16.27	17.59	19.20	17.40	19.17	20.52	21.19
May	14.66	14.25	15.33	16.11	17.20	17.36	18.60	18.55	18.53	20.14	21.23	22.51
June	14.58	14.66	15.60	16.58	17.11	16.60	18.71	18.92	18.91	20.18	20.74	22.24
July	14.38	14.39	15.26	15.38	16.86	16.60	18.33	18.07	17.92	19.07	20.54	22.15
August	14.18	13.90	15.48	16.10	17.49	17.00	18.05	18.15	18.33	19.85	21.02	21.78
September	14.08	14.14	15.76	15.58	16.37	16.33	17.57	17.90	18.16	18.79	19.27	21.31
October	14.95	14.66	15.68	16.13	16.95	17.36	19.10	18.65	18.77	20.50	21.53	22.61
November	13.96	14.53	15.75	16.49	17.13	17.04	17.64	18.41	19.28	20.83	21.56	21.70
December	16.44	17.87	19.12	16.38	19.84	21.17	21.45	22.15	22.87	24.09	25.10	27.72

Using x_1 = year (coded 1 to 12) and x_2 = month (coded 1 to 12), the model is specified:

$$y = \beta_{00} + \beta_{10}x_1 + \beta_{20}x_1^2 + \beta_{01}x_2 + \beta_{02}x_2^2 + \beta_{03}x_2^3$$
$$+ \beta_{11}x_1x_2 + \beta_{12}x_1x_2^2 + \beta_{13}x_1x_2^3 + \varepsilon$$

The estimated equation is

$$\hat{\mu}_x = 10.43 + 0.1214x_1 + 0.02106x_1^2 + 1.584x_2 - 0.2381x_2^2$$
$$+ 0.01169x_2^3 + 0.1479x_1x_2 - 0.02652x_1x_2^2 + 0.001463x_1x_2^3$$

The equation is highly significant [$F_{(8,135)}$ = 130.31], R^2 = 0.913, and all coefficients except β_{10} are significant at the 0.05 level.

The effect of years (x_1) is clearly an accelerating upward trend. The following rearrangement of terms provides the description of the monthly trend:

$$\hat{\mu}_x = 10.43 + 0.1214x_1 + 0.02106x_1^2 + (1.584 + 0.1479x_1)x_2$$
$$+ (-0.2381 - 0.02652x_1)x_2^2 + (0.01169 + 001463x_1)x_2^3$$

The interaction coefficients (β_{11}, β_{12}, β_{13}) are obviously tending to increase the magnitudes of the linear, quadratic, and cubic trends associated with months. This can be seen by constructing the equations for 1953(x_1 = 1), 1959(x_1 = 7), and 1964(x_1 = 12):

1953(x_1 = 1)

$$\hat{\mu}_x = 10.57 + 1.732x_2 - 0.2646x_2^2 + 0.1493x_2^3 + \cdots$$

1959(x_1 = 7)

$$\hat{\mu}_x = 12.31 + 2.620x_2 - 0.4237x_2^2 + 0.1581x_2^3 + \cdots$$

1964(x_1 = 12)

$$\hat{\mu}_x = 14.92 + 3.359x_2 - 0.5563x_2^2 + 0.1655x_2^3 + \cdots$$

The monthly trends resulting from these equations are illustrated in Figure 6.3. It is not difficult to see how the effects of the three polynomial terms (see Figure 6.1) increase from 1953 to 1964.

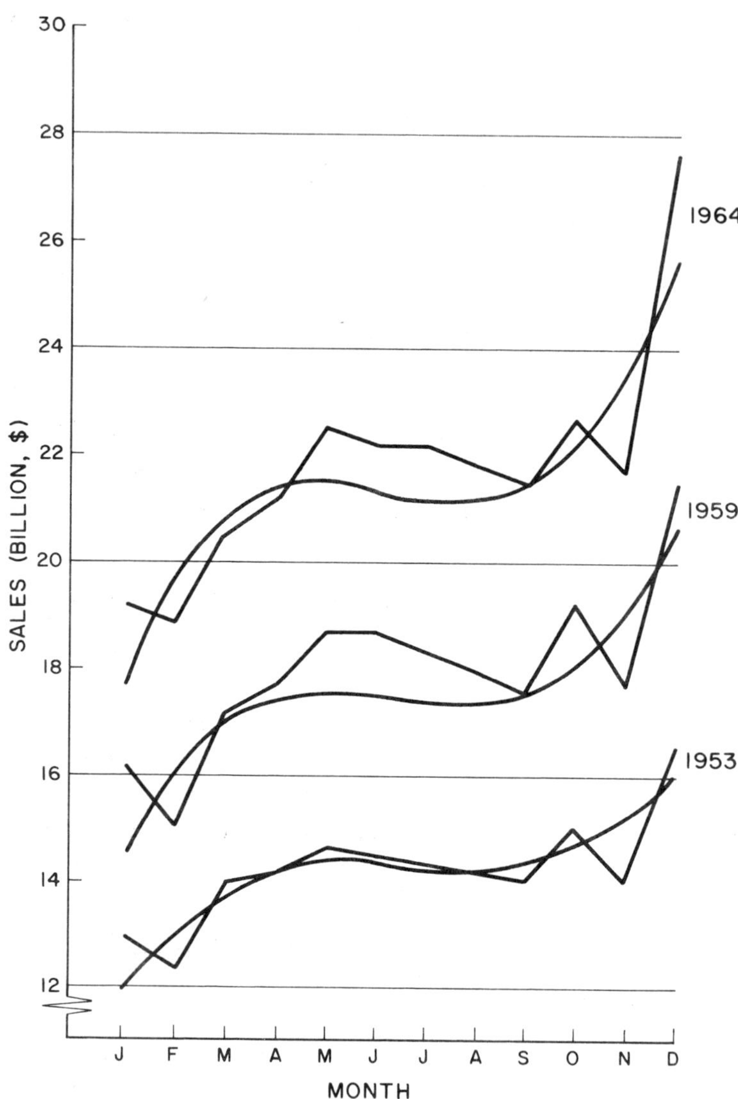

Figure 6.3 Estimated response surfaces.

6.5 ANALYSIS OF RESPONSE SURFACE EXPERIMENTS

Polynomial regressions are often used in response surface analyses, which attempt to obtain estimates of optimum responses, or yields, from various combinations of inputs. Common examples of such analyses include crop yield responses to fertilizer inputs or yields of chemical reactions as functions of the settings of controls of equipment. Such analyses are most effective when based on data obtained from designed experiments, and a number of special designs have been developed for such purposes. A discussion of the construction and proper usage of such designs is beyond the scope of this book; this section discusses the analysis of such data as it relates to regression analysis. More complete discussions may be found in chapters or books on factorial experiments [e.g., Snedecor and Cochran (1967, Chapter 12)].

A very popular experimental design for response surface analyses consists of performing a complete factorial experiment in which all possible combinations of treatment levels are observed, preferably with replication to provide an appropriate error. As an example, consider a 4 × 3 factorial experiment: 12 treatment combinations, consisting of four levels of factor A and three levels of factor C, with two replications in a completely randomized design. The appropriate analysis of variance format for this experiment is given in Table 6.7.

TABLE 6.7 Factorial Analysis Partitioning

Source	d.f.
Treatments	11
A	3
C	2
AC	6
Error	12

Recall from Chapter 4 that k - 1 regression coefficients may be estimated to correspond to the k - 1 single degree of freedom sum of squares describing the variation among k groups or treatments. Assume that the two factors in the 4 × 3 factorial experiment correspond to the quantitative factors (e.g., pounds of fertilizer components) labeled x_1 and x_2. The 12 treatment combinations then correspond to 12 combinations of the four values of x_1 and three values of x_2. The 11 regression coefficients describing the variation among these treatments correspond to the 11 terms (plus the intercept) of the polynomial equation

$$y = \sum_{i=0}^{3} \sum_{j=0}^{2} \beta_{ij} x_1^{\,i} x_2^{\,j} + \varepsilon$$

The equation contains all possible polynomial terms involving exponents of x_1 and x_2 of degree one less than the number of factor levels. Furthermore, there is a correspondence between sets of terms in the polynomial equation and the individual partitions of the sum of squares; this is given in Table 6.8. Note that if fewer than 11 terms are specified in the polynomial model, the difference between the treatment and regression sums of squares may be used in a lack-of-fit test for adequacy of the model (Section 4.2).

TABLE 6.8 Correspondence between Polynomial and Factorial Analysis

Source	d.f.	Corresponding polynomial terms
A	3	$\beta_{10}x_1$, $\beta_{20}x_1^2$, $\beta_{30}x_1^3$
C	2	$\beta_{01}x_2$, $\beta_{02}x_2^2$
AC	6	$\beta_{11}x_1x_2$, $\beta_{12}x_1x_2^2$, $\beta_{21}x_1^2x_2$, $\beta_{22}x_1^2x_2^2$, $\beta_{31}x_1^3x_2$, $\beta_{32}x_1^3x_2^2$

If the levels of factors in a factorial experiment are equally spaced, orthogonal polynomials can be used to facilitate the analysis.

TABLE 6.9 Data for Response Surface

Factor A level	Factor C level $x_2 = 2$	$x_2 = 4$	$x_2 = 6$	Totals
$x_1 = 10$	75	61	37	
	69	52	58	
	(144)	(113)	(95)	352
$x_1 = 20$	73	40	36	
	55	57	31	
	(128)	(97)	(67)	292
$x_1 = 30$	89	58	89	
	74	82	77	
	(163)	(140)	(166)	469
$x_1 = 40$	63	71	83	
	83	68	92	
	(146)	(139)	(175)	460
Totals	581	489	503	1573

This is illustrated with an example. Table 6.9 gives hypothetical data for a 4 × 3 factorial with two replications; factor A has four levels corresponding to x_1-values 10, 20, 30, and 40, while factor C has three levels with x_2-values 2, 4, and 6. Cell totals of the two (replicated) observations are given in parentheses, while main effect totals and the grand total are given in margins. The appropriate factorial analysis of variance is given in Table 6.10; only the A main effect is statistically significant. Normally it would be logical to examine only polynomials in x_1 (the coefficients associated with the A effect), but for purposes of illustration, a polynomial model is specified to include terms in both x_1 and x_2:

$$y = \beta_{00} + \beta_{10}x_1 + \beta_{20}x_1^2 + \beta_{30}x_1^3 + \beta_{01}x_2 + \beta_{11}x_1x_2 + \beta_{21}x_1^2x_2 + \beta_{31}x_1^3x_2 + \varepsilon$$

TABLE 6.10 Factorial Analysis of Variance

Source	d.f.	SS	MS	F
A	3	3691.12	1230.38	11.23[a]
C	2	614.33	307.17	2.80
AC	6	1497.00	249.50	2.28
Error	12	1315.00	109.62	

[a]Statistically significant at the 0.05 level.

Table 6.11 shows the most commonly used format for performing the indicated analysis. Briefly, the details of the procedure are as follows:

1. The integers in the body of the table are the orthogonal polynomial values. Specifically, for the A and C main effects, the appropriate orthogonal polynomial values for $n = 4$ and $n = 3$ are repeated as needed to correspond to levels of A and C, respectively; the AC interaction polynomial values are simply products of the corresponding main effect polynomial values
2. The polynomial values are used as independent variables as in the one-variable polynomial regression. Since in this example the dependent variable values are totals (of two observations), the denominator in expressions for the coefficients and sums of squares becomes $n \sum z^2$, with $n = 2$

The sums of squares obtained from the polynomial regression are then combined with those of the factorial analysis of variance as given in Table 6.12. The linear and cubic effects of A and the linear by linear interaction are statistically significant; the small lack-of-fit mean square indicates that the model used is adequate. However, since the cubic term in A is to be used, the quadratic must also be included.

TABLE 6.11 Orthogonal Polynomials for Factorial Analysis

Level of A:	10			20			30			40						
Level of C:	2	4	6	2	4	6	2	4	6	2	4	6	$\sum zy$	$n \sum z^2$	$\frac{\sum zy}{n \sum z^2}$	$\frac{(\sum zy)^2}{n \sum z^2}$
Cell total:	144	113	95	128	97	67	163	140	166	146	139	175			$(\hat{\alpha}_i)$	$SS(\hat{\alpha}_i)$
Polynomials:																
A linear	-3	-3	-3	-1	-1	-1	1	1	1	3	3	3	501	120	4.175	2091.67
A quadratic	1	1	1	-1	-1	-1	-1	-1	-1	1	1	1	51	24	2.125	108.38
A cubic	-1	-1	-1	3	3	3	-3	-3	-3	1	1	1	-423	120	-3.525	1491.08
C linear	-1	0	1	-1	0	1	-1	0	1	-1	0	1	-78	16	-4.875	380.25
$A_L \times C_L$:	3	0	-3	1	0	-1	-1	0	1	-3	0	3	298	80	3.725	1110.05
$A_Q \times C_L$:	-1	0	1	1	0	-1	1	0	-1	-1	0	1	38	16	2.375	90.25
$A_C \times C_L$:	1	0	-1	-3	0	3	3	0	-3	-1	0	1	-114	80	-1.425	162.45

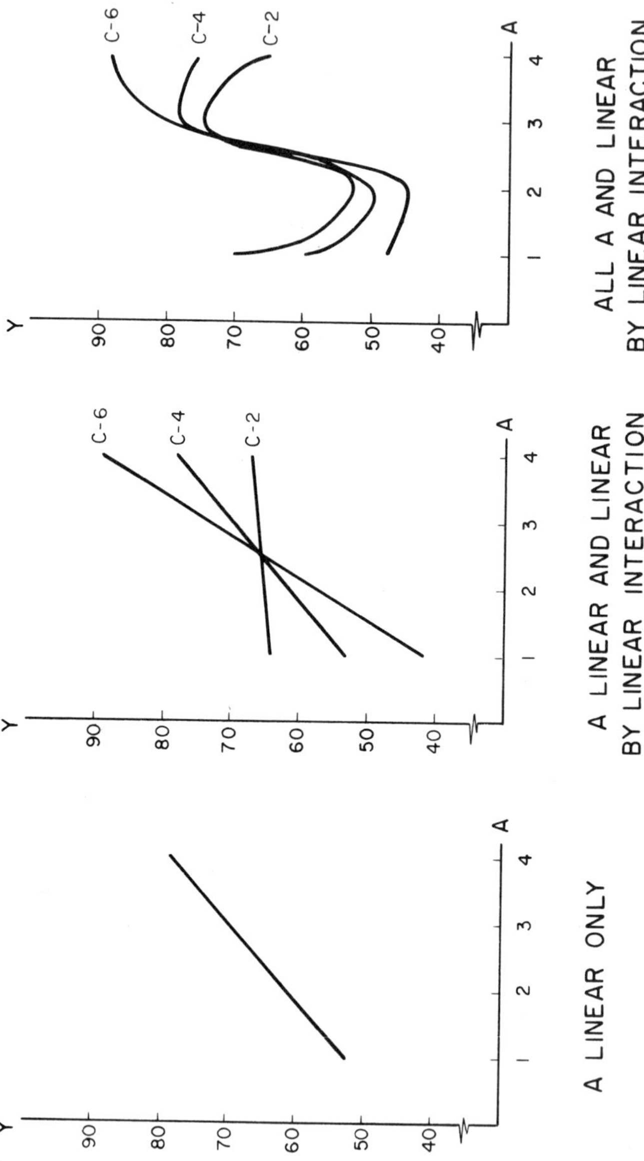

Figure 6.4 Response surfaces for example.

TABLE 6.12 Combined Analysis

Source	d.f.	SS	MS	F
A (Main effect)	3	3691.12	1230.38	11.23[a]
Linear	1	2091.67	2091.67	19.08[a]
Quadratic	1	108.37	108.37	0.99
Cubic	1	1491.07	1491.07	13.60[a]
C (Main effect)	2	614.33	307.17	2.80
Linear	1	380.25	380.25	3.47
Lack of fit	1	234.08	234.08	2.14
AC (Interaction)	6	1497.00	249.50	2.28
A LIN × C LIN	1	1110.05	1110.05	10.12[a]
A QUAD × C LIN	1	90.25	90.25	0.82
A CUB × C LIN	1	162.45	162.45	1.48
Lack of fit	3	134.25	44.75	0.41
Error	12	1315.50	109.625	

[a]Statistically significant at 0.01 level.

Figure 6.4 provides an illustration of the components of the estimated response surfaces. The first graph shows the estimated average linear (only) response to A, the second shows how this effect is modified by the existence of the linear by linear interaction, and the third shows the final addition of the (small) quadratic U-shaped curve and the cubic "hook" (see Figure 6.1).

6.6 NONLINEAR REGRESSION

Most nonlinear regression problems (nonlinear in the parameters) cannot be solved by the procedures presented in this chapter. The main difficulty with such problems is that the direct application of the least-squares principle requires solutions to systems of nonlinear equations. Consider, for example, the simple "decay" curve, which (among other things) is used to describe the decay of a radioactive substance:

$$y = \beta_0 + \beta_1 e^{-\beta_2 x} + \varepsilon$$

where

y = weight of the substance at time x

x = time

In this function the quantity $\beta_0 + \beta_1$ is the initial weight (when $x = 0$) and β_0 is the ultimate weight (as x tends to infinity) of the nondecaying portion of the material. The residual sum of squares to be minimized is

$$\text{SSE} = \sum(y - \beta_0 - \beta_1 e^{-\beta_2 x})^2$$

in which the summation is over the sample values of x and y. It is not difficult to take the partial derivatives of this quantity with respect to β_0, β_1, β_2, respectively. However, equating the derivatives to zero produces a set of normal equations for which there is no *closed* solution, that is, no solution for which a general (algebraic) form can be given. Therefore the system must be solved by numerical methods. Furthermore, there exist no closed expressions relating to the partitioning of the sums of squares and variances of coefficients.

Most of the research in nonlinear regression falls into three major categories:

1. Approximate methods
2. Simpler and more effective numerical procedures for solving the normal equations
3. Procedures for finding estimates of variances or confidence limits for the coefficients of nonlinear equations

Discussions of such procedures are beyond the scope of this book. Some of the methods are implemented in the more complete statistical computer software packages, whose documentation usually contains a selected set of references.

There exists, however, one simple procedure which is useful for a special case of nonlinear regression problems. It has already been indicated that the basic theory underlying linear regression holds for any function linear in the parameters and that there is no restriction on the nature of the independent variables. An interesting class of problems arises when transformations are performed on the *dependent* variable. For example, assume the model

$$\sqrt{y} = \beta_0 + \beta_1 x_1 + \varepsilon$$

and estimate the parameters β_0 and β_1 by least squares. Note, however, that the corresponding model with y as the dependent variable is

$$y = \beta_0^2 + \beta_1^2 x_1^2 + 2\beta_0\beta_1 x_1 + 2\beta_0\varepsilon + 2\beta_1 x_1 \varepsilon + \varepsilon^2$$

a truly nonlinear model which fails to fulfill the usual assumptions underlying linear regression models. The basic difficulty arises from the fact that the model is not strictly additive in the error term: it is not of the form

$$y = f(\underline{x}) + \varepsilon$$

and hence the least-squares estimate obtained using the transformed dependent variable is *not* a least-squares estimate of μ_x.

Nevertheless, there are some special cases of this type which have been widely used. In particular, the logarithmic model is of great interest. Define a model linear in the logarithms[†] of *both* independent and dependent variables:

$$\log y = \beta_0 + \beta_1 \log x_1 + \beta_2 \log x_2 + \cdots + \beta_m \log x_m + \varepsilon$$

[†]Logarithms to base e are usually used; the use of base 10 provides equivalent results; β_0 and s^2 will be different, but all other results given here will be the same.

and estimate the coefficients by usual methods. In terms of the original variables, the model is

$$y = e^{\beta_0} x_1^{\beta_1} x_2^{\beta_2} \cdots x_m^{\beta_m} e^{\varepsilon}$$

This model has some very important uses. The resulting coefficients are *elasticities*, that is, each β_i reflects the *percent* change in y associated with a 1% change in the corresponding x_i. Further the quantity $\sum_{i=1}^{m} \beta_i$ reflects the percent change in y associated with a 1% change in *each* x_i. If the model represents a production function, relating the output of a production process to quantities of inputs to the process, economists call it a *Cobb-Douglas production function*. It is also sometimes called a *learning curve*.

This equation is useful only if values of all variables are strictly positive, since the model specifies y = 0 if *any* x = 0 (also, log 0 = −∞). Occasionally, however, it may be used when some values of zero occur, in which case a small positive constant may be added to all values before logarithms are taken.

Note further that the error term is *multiplicative*. This means that the coefficients estimated by the use of least squares on logarithms do not provide the least-squares estimates of the multiplicative function. They are what would be obtained by generating normally distributed random errors, using these as exponents and multiplying resulting μ_x-values by this quantity. Note that if the random error is zero, then $e^0 = 1$ and the y-value is equal to μ_x; a negative random error produces a multiplicative factor of less than one, etc. The least-squares procedure in logarithms is then actually some sort of "minimum product of errors" procedure for the exponential model. The above procedure is not, however, necessarily incorrect, since the random errors for situations where the exponential model is appropriate are often actually multiplicative in nature.

Occasionally in the analysis of time series data it is desired to use first differences of logarithms; this procedure implies a multiplicative model in *ratios* of adjacent observations: y_t/y_{t-1}.

Since multiplicative errors are conceptually difficult to interpret and since it is sometimes desired to make a comparison of the fit of a multiplicative and, say, a polynomial model, it is desirable to obtain an estimate of the equivalent additive error variance. The most obvious procedure is to obtain the estimated $(\widehat{\log \mu_x})$ values, take antilogarithms (exponentiate), call these $\tilde{\mu}_x = \log^{-1}(\widehat{\log \mu_x})$, then compute

$$SSE = \sum(y - \tilde{\mu}_x)^2$$

Actually this is a biased estimate of the residual sum of squares, although the bias is small when R^2 is near unity. Brownlee (1960, Section 3.2) provides a method which reduces this bias. Instead of using $\tilde{\mu}_x$ in the formula for SSE, compute

$$\tilde{\mu}_x^* = \log^{-1}[(\widehat{\log \mu_x}) + k(\text{MSE, using logs})]$$

where k = 0.5. (If logarithms to base 10 have been used, k = 1.15.) Note that neither $\Sigma(y - \tilde{\mu}_x)$ nor $\Sigma(y - \tilde{\mu}_x^*)$ is usually zero.

An example of a model suitable for this type of analysis is afforded by the example given at the beginning of this chapter (Table 6.1). The correct model is $\mu_x = x^{0.5}$, which transforms to $\log \mu_x = 0.05 \log x$; the true error is *not*, however, multiplicative. Using the logarithmic model on the first seven observations results in the estimated equation

$$\log y = -0.1954 + 0.5508 \log x$$

which is quite close to the true model. The residual standard deviation of the logarithmic equation is 0.1395. The "plus or minus one standard deviation" equivalent for the multiplicative model is the multiplication by $e^{\pm 0.1395}$, which gives multiplicative factors of 1.15 and 0.87, respectively, or a multiplicative error of 13 to 15%.

The estimated values $\tilde{\mu}$ are given in Table 6.13 and are all seen to be lower than the true model values; the adjusted estimates $\tilde{\mu}^*$ provide smaller residuals and, since the *model* is correct, provide closer extrapolated values than did the polynomial approximation.

TABLE 6.13 Exponential Model Estimation

x	μ	$\tilde{\mu}$	$\tilde{\mu} - \mu$	$\tilde{\mu}^*$	$\tilde{\mu}^* - \mu$
1	1.000	0.823	-0.177	0.882	-0.118
2	1.414	1.205	-0.209	1.292	-0.122
3	1.732	1.506	-0.226	1.615	-0.117
4	2.000	1.765	-0.235	1.892	-0.108
5	2.236	1.996	-0.240	2.140	-0.096
6	2.449	2.206	-0.243	2.366	-0.083
7	2.646	2.402	-0.244	2.575	-0.071
8	2.828	2.585[a]	-0.243[a]	2.772[a]	-0.056[a]
9	3.000	2.759[a]	-0.241[a]	2.958[a]	-0.042[a]
10	3.162	2.823[a]	-0.239[a]	3.135[a]	-0.027[a]

[a]Extrapolated values.

6.7 PROBLEMS

1. The x-ray picture of a heel bone was divided into 10 segments of equal length. Four readings of the x-ray were taken for the *densities* of the bone for each segment. Data are given in Table 6.14. Do the appropriate analysis of the data, including a polynomial equation describing the pattern of the density along the bone, and the analysis for adequacy of the model

2. It is desired to study the effect of latitude and longitude on temperature in the eastern half of the United States. We have average weekly temperatures for the first 25 weeks of 1967 for the 14 cities given on the map for problem 2 of Chapter 4. (The data are not reproduced here.) Using latitudes and longitudes as coded on the map, a polynomial regression model is proposed as follows:

TABLE 6.14 Heel-bone Data

	Reading			
Segment	1	2	3	4
1	549	554	542	545
2	840	851	843	841
3	1095	1096	1105	1096
4	1284	1282	1303	1302
5	1353	1350	1352	1335
6	1339	1334	1336	1333
7	1280	1277	1202	1281
8	1284	1280	1300	1285
9	1302	1320	1293	1311
10	1281	1293	1309	1285

$$y = \beta_{000} + \beta_{100}x_1 + \beta_{010}x_2 + \beta_{001}x_3$$
$$+ \beta_{002}x_3^2 + \beta_{003}x_3^3 + \beta_{101}x_1x_3 + \beta_{011}x_2x_3 + \varepsilon$$

where

x_1 = latitude (LAT)

x_2 = longitude (LONG)

x_3 = week (W)

Some results are given in Table 6.15. Complete the analysis; interpret results *fully*. What climatological fact do the results support?

3. Given the artificial data (Table 6.16) from a 2 × 5 factorial experiment with no replications:

   (a) Do the appropriate analysis of variance and obtain sums of squares for

A linear

B linear

(A linear) × (B linear)

Interpret the results fully.

TABLE 6.15 Problem 2, Some Results

Variable	Mean	Coef.	t-value
Intercept = β_0	--	-11.103	--
LAT	2.5	15.138	21.156
LONG	2.357	3.1289	4.608
W	13	0.51991	0.894
W^2	221	0.19246	3.912
W^3	4225	-0.0039173	-3.145
W × LAT	32.5	-0.32677	-6.789
W × LONG	30.64	-0.23712	-5.192

Source	d.f.	SS
Total	349	125,205.7
Regression	7	109,375.6
Residual	342	15,830.1

TABLE 6.16 Artificial Data

Factor C	Factor A (x_1)					
(x_2)	1	2	3	4	5	Total
10	3	4	5	9	14	35
20	5	6	9	7	8	35
Total	8	10	14	16	22	70

(b) Perform the regression analysis using the model

$$y = \beta_{00} + \beta_{10}x_1 + \beta_{01}x_2 + \beta_{11}x_1x_2$$

(c) Compare results of (a) and (b).

4. We are investigating the permanent stress (deformation) of asphalt pavements due to different "aggregates" in their mix and of various levels of strain (in pounds per square inch, psi). The data are given in Table 6.17. Perform the appropriate analysis and interpret results.

5. Refer to the feeder cattle placement problem in Chapter 2 (problem 4). The use of the (natural) logarithmic transformation produces the following results:

$$(X'X)^{-1} = \begin{bmatrix} 6.1932 & -6.4767 & 2.4737 \\ -6.4767 & 10.7358 & -1.0013 \\ 2.4737 & -1.0013 & 3.6844 \end{bmatrix} \qquad \hat{\underline{\beta}} = \begin{bmatrix} -1.9820 \\ 2.1951 \\ -2.4962 \end{bmatrix}$$

$$\hat{\beta}_0 = 7.2866 \qquad \hat{\underline{\beta}}'X'\underline{y} = 2.0784 \qquad \underline{y}'\underline{y} = 4.2076$$

Complete the analysis and interpret results.

TABLE 6.17 Permanent Stress on Asphalt Mixtures

Strain (psi)	Aggregate: Gravel	Light weight	Limestone
50	0.73	0.45	0.32
	0.77	0.55	0.33
100	4.4	1.67	1.39
	4.2	1.12	2.01
150	10.36	1.88	3.38
	10.12	2.06	4.04

6. An experiment is conducted to determine the effect on various responses (e.g., temperature) of the innoculation of a nonlethal virus. Two calves were chosen for the experiment and closely observed for 4 days.

   On the morning of the fifth day the calves were innoculated, and the observation continued for 4 additional days afterwards. The responses consist of a number of tests, some requiring withdrawal of blood and other fluids. The data for one of the responses (called y) for the 9 days of observations for the two calves is given in Table 6.18.

   A statistician, trying to be helpful, suggests a regression analysis using some artificial variables called z_1, z_2, z_3, and z_4; the values are given in Table 6.18.

   Some calculations:

$$\sum(y - \bar{y})^2 = 120.503$$
$$\sum z_1 y = 4.3$$
$$\sum z_2 y = 4.0$$
$$\sum z_3 y = 31.7$$
$$\sum z_4 y = 96.5$$

   (a) What are these variables supposed to show?
   (b) Complete the analysis with relevant estimates and tests. *Interpret.*

   After seeing these results, the statistician says you should also do an analysis of variance using days and calves as sources of variation. The calculations give

   SS between days = 102.728
   SS between calves = 5.014

(c) Complete the analysis and incorporate with results of part (b). Comment on what this analysis shows over and above the results of part (b).

(d) Can you recommend any other possible additions to the analysis?

TABLE 6.18 Experimental Data

Calf	Day	Condition	z_1	z_2	z_3	z_4	y
1	1	Pre	-1	-3	0	-1	5.5
	2	Pre	-1	-1	0	-1	3.8
	3	Pre	-1	1	0	-1	5.8
	4	Pre	-1	3	0	-1	6.1
	5	Shot	0	0	0	8	11.2
	6	Post	1	0	-3	-1	8.9
	7	Post	1	0	-1	-1	3.8
	8	Post	1	0	1	-1	5.1
	9	Post	1	0	3	-1	3.9
2	1	Pre	-1	-3	0	-1	6.7
	2	Pre	-1	-1	0	-1	4.8
	3	Pre	-1	1	0	-1	5.6
	4	Pre	-1	3	0	-1	6.5
	5	Shot	0	0	0	8	12.6
	6	Post	1	0	-3	-1	9.8
	7	Post	1	0	-1	-1	8.7
	8	Post	1	0	1	-1	3.3
	9	Post	1	0	3	-1	5.6
$\sum(\ )$			0	0	0	0	117.7
$\sum(\ )^2$			16	40	40	144	890.13

Chapter 7

GENERAL LINEAR MODELS

7.1 INTRODUCTION

Previous chapters have discussed general multiple (linear) regression analysis based on the model

$$y = \beta_0 + \beta_1 x_1 + \beta_2 x_2 + \cdots + \beta_m x_m + \varepsilon$$

The method of least squares has been used to obtain estimates of the β_i, and partitions of sums of squares have led to analysis of variance tests of hypotheses regarding individual coefficients, as well as subsets or linear functions of coefficients.

It was noted in Chapter 2 that there is no restriction on the nature of the independent variables. Later in that chapter a dummy variable, whose value is always unity, was used as a device to estimate the y-intercept β_0. In Chapter 6, artificially coded orthogonal polynomials were used to estimate a polynomial relationship. In this chapter, procedures are developed for generating artificial variables which, when used as independent variables in a regression model, produce results most commonly associated with the analysis of variance.

As an example of the type of variables to be used, consider the following situation:

There are two groups (samples) of size n_1 and n_2 of observed values, y, with means $\bar{y}_1$ and $\bar{y}_2$, respectively. Associated with these observations is an independent variable, x, which takes the value zero for any observation in group 1 and unity for any observation in group 2. We perform a regression analysis according to the model

$$y = \beta_0 + \beta_1 x + \varepsilon$$

Denoting $\bar{x}$, $\bar{y}$, and n as the means and sample sizes of the entire data set, the following equivalences for quantities involved in the regression analysis can be derived:

1. $\sum(x - \bar{x})^2 = n_1 n_2/(n_1 + n_2)$
2. $\sum(x - \bar{x})(y - \bar{y}) = n_1 n_2(\bar{y}_2 - \bar{y}_1)/(n_1 + n_2)$
3. $\sum(y - \bar{y})^2 = \sum y^2 - (n_1\bar{y}_1 + n_2\bar{y}_2)^2/(n_1 + n_2)$

These are used to estimate [formula (2.14)]

$$\hat{\beta} = (\bar{y}_1 - \bar{y}_2)$$

which is the difference between the group means. To test the hypothesis $\beta = 0$, compute [formula (3.3)]

$$\text{SSR} = \frac{n_1 n_2(\bar{y}_2 - \bar{y}_1)^2}{n_1 + n_2}$$

hence,

$$\text{SSE} = \sum y^2 - n_1\bar{y}_1{}^2 - n_2\bar{y}_2{}^2$$

which is recognized as the pooled (within-group) sum of squares. Finally, the test statistic (Table 3.3),

$$F = \frac{(\bar{y}_1 - \bar{y}_2)^2}{\dfrac{\text{SSE}}{n_1 + n_2 - 2}\;\dfrac{n_1 + n_2}{n_1 n_2}} \qquad (\text{d.f.} = 1,\ n - 2)$$

This is also the square of the familiar "pooled t test" or the F test for a two-treatment completely randomized experiment.

The procedures presented in this chapter are concerned with extensions of this type of regression analysis for performing the analysis of variance for any given situation. The associated procedures are considerably more cumbersome than the usual ANOVA calculations; however, they can be used in situations where the usual formulas fail. The methodology variously goes under the names of least squares, fitting constants, linear models, or linear hypotheses. Complete discussions of this topic can be found in Graybill (1976) or Searle (1971).

7.2 EXAMPLE: TWO-FACTOR BALANCED EXPERIMENT

Consider an artificial example of a two-factor experiment with a = 3 levels of a treatment called the A *treatment* and c = 4 levels of the C *treatment*. Each of the 3 × 4 = 12 treatment combinations is replicated n = 2 times. The data are given in Table 7.1. The linear model used as the basis for an analysis of variance for the data is

$$y_{ijk} = \mu + \alpha_i + \gamma_j + (\alpha\gamma)_{ij} + \varepsilon_{ijk} \tag{7.1}$$

where

y_{ijk} = observed response of the k-th replication at the i-th level of A and j-th level of C treatment combination

μ = population mean

α_i = effect of the i-th A treatment (i = 1, 2, 3)

γ_j = effect of j-th C treatment (j = 1, 2, 3, 4)

$(\alpha\gamma)_{ij}$ = the interaction effect of the i - j treatment combination

ε_{ijk} = the random error, normally distributed with mean zero and variance σ^2 (k = 1, 2)

An additional specification usually imposed is that various sums of treatment effects are zero:

$$\sum \alpha_i = \sum \gamma_j = \sum_i (\alpha\gamma)_{ij} = \sum_j (\alpha\gamma)_{ij} = \sum_i \sum_j (\alpha\gamma)_{ij} = 0 \tag{7.2}$$

TABLE 7.1 Two-Factor Experiment; Equal Cell Frequencies

	C treatment				
A treatment	1	2	3	4	Total
1	3	5	10	11	56
	5	9	8	5	
2	6	6	9	10	72
	10	12	13	6	
3	4	7	6	4	40
	2	3	8	6	
Total	30	42	54	42	168

these are imposed in order to avoid ambiguities in definitions (see the discussion in Section 7.3).

The recommended analysis of these data is the familiar analysis of variance [see, e.g., Snedecor and Cochran (1967, Chapters 11-13)]. First obtain the total sum of squares:

$$\text{SS total} = \sum\sum\sum y_{ijk}^2 - \frac{Y_{\cdot\cdot\cdot}^2}{acn}$$

$$= 1382 - 1176 = 206$$

The sum of squares among the 12 treatment combinations (cells) is

$$\text{SS cells} = \frac{\sum Y_{ij\cdot}^2}{n} - \frac{Y_{\cdot\cdot\cdot}^2}{acn}$$

$$= 1296 - 1176 = 120$$

The error sum of squares is obtained by subtraction:

$$\text{SS error} = 206 - 120 - 86$$

The sum of squares due to the A treatment is

$$\text{SSA} = \frac{\sum Y_{i\cdot\cdot}^2}{nc} - \frac{Y_{\cdot\cdot\cdot}^2}{acn}$$

$$= 1240 - 1176 = 64$$

the sums of squares due to the C treatment is

$$SSC = \frac{\sum Y^2_{\cdot j \cdot}}{an} - \frac{Y^2_{\cdots}}{acn}$$

$$= 1224 - 1176 = 48$$

and the interaction sum of squares is obtained by subtraction:

$$\text{SS interaction} = \text{SS cells} - \text{SSA} - \text{SSC}$$
$$= 120 - 64 - 48 = 8$$

The above results are summarized in the ANOVA table (Table 7.2), which indicates a (5%) statistically significant treatment effect for the A treatment (assuming *fixed-effects* model). The fixed-effects model is assumed throughout this chapter [see, e.g., Ostle and Mensing (1975, p. 282)].

The estimates of the model parameters are

$$\hat{\mu} = \bar{y}_{\cdots} = 7$$

$$\hat{\alpha}_1 = \bar{y}_{1\cdot\cdot} - \bar{y}_{\cdots} = 7.0 - 7.0 = 0.0$$

$$\hat{\alpha}_2 = \bar{y}_{2\cdot\cdot} - \bar{y}_{\cdots} = 9.0 - 7.0 = 2.0$$

....

$$\hat{\gamma}_4 = \bar{y}_{\cdot 4 \cdot} - \bar{y}_{\cdots} = 7.0 - 7.0 = 0.0$$

TABLE 7.2 Analysis of Variance

Source	d.f.	SS	MS	F
A treatment	2	64	32	4.47[a]
C treatment	3	48	16	2.23
Interaction (A × C)	6	8	1.33	0.19
Error (within)	12	86	7.17	
Total	23	206		

[a]Statistically significant at the 0.05 level.

The interaction effects are estimated similarly, and finally the estimated variance

$$s^2 = \text{MS error} = 7.17$$

We now consider estimation and hypothesis testing for this model using regression methods. The model is simplified to

$$y_{ijk} = \mu + \alpha_i + \gamma_j + \varepsilon'_{ijk} \tag{7.3}$$

where ε'_{ijk} contains interaction *and* error. The true error will be computed directly from the within-cell sum of squares and the interaction obtained by subtraction. Interaction can be included in this model but is excluded here for simplicity (see Section 7.5).

An inspection of this model shows one major departure from the usual regression model: there are no independent variables. This is, however, remedied by the device of introducing into the model several sets of independent dummy variables as follows:

$$y_{ijk} = \mu x + \sum_{i=1}^{3} \alpha_i v_i + \sum_{j=1}^{4} \gamma_j w_j + \varepsilon'_{ijk} \tag{7.4}$$

where

$$x = 1 \text{ always}^{\dagger}$$

$$v_i = \begin{cases} 1 & \text{for any observation in the i-th level of A} \\ 0 & \text{otherwise} \end{cases}$$

$$w_j = \begin{cases} 1 & \text{for any observation in the j-th level of C} \\ 0 & \text{otherwise} \end{cases}$$

This device provides a regression model with a total of eight independent variables: 1 x variable, 3 v variables, and 4 w variables,

---

†This can be left off and, as is done with the β_0 term (Chapter 2), be handled by correcting other observations for the mean. The usual method is, however, to use this dummy variable.

TABLE 7.3 The Observation (Design) Matrix

A	C	Rep	x	v_1	v_2	v_3	w_1	w_2	w_3	w_4	y
1	1	1	1	1			1				3
		2	1	1			1				5
	2	1	1	1				1			5
		2	1	1				1			9
	3	1	1	1					1		10
		2	1	1					1		8
	4	1	1	1						1	11
		2	1	1						1	5
2	1	1	1		1		1				6
		2	1		1		1				10
	2	1	1		1			1			6
		2	1		1			1			12
	3	1	1		1				1		9
		2	1		1				1		13
	4	1	1		1					1	10
		2	1		1					1	6
3	1	1	1			1	1				4
		2	1			1	1				2
	2	1	1			1	1	1			7
		2	1			1		1			3
	3	1	1			1			1		6
		2	1			1			1		8
	4	1	1			1				1	4
		2	1			1				1	6

which, for any specific observation, exactly reproduces the original model.

The values of the x, v, and w variables comprise the X matrix and the y-values the y-vector required for performing a regression analysis; these matrices are given in Table 7.3. For this type of analysis the X matrix is often called the *design matrix*, as it specifies the arrangement of treatment combinations, i.e., the *design* of the data.

The next step is to obtain X'X, the matrix sums of squares and cross products of the independent variables, and $X'\underline{y}$, the vector of cross products with the dependent variable. These are given in Table 7.4; the parameters corresponding to the dummy variables are given in the row headings. Note that these are uncorrected X'X and $X'\underline{y}$ matrices since the y-intercept μ is included in the model.

TABLE 7.4 Sums of Squares and Cross Products

		(X'X)								
Param.	Var.	x	v_1	v_2	v_3	w_1	w_2	w_3	w_4	$X'\underline{y}$
μ	x	24	8	8	8	6	6	6	6	168
α_1	v_1	8	8			2	2	2	2	56
α_2	v_2	8		8		2	2	2	2	72
α_3	v_3	8			8	2	2	2	2	40
γ_1	w_1	6	2	2	2	6				30
γ_2	w_2	6	2	2	2		6			42
γ_3	w_3	6	2	2	2			6		54
γ_4	w_4	6	2	2	2				6	42

Since the elements of the X matrix consist exclusively of ones and zeros, all sums of squares and cross products are simply frequencies of the occurrences of ones. For example,

$\sum x^2$ = the sum of ones over all observations; hence, it is the total sample size, which is 24 in this example

$\sum v_1^2$ = the sum of ones when v_1 is one; hence, it is the number of observations in the first level of A, which is 8 in this example

$\sum xv_1$ = the sum of ones in the x column when v_1 is equal to one; hence, it is the number of observations in the first level of A, which is 8 in this example

$\sum v_1v_2$ = zero since v_1 and v_2 (two levels of A) cannot occur simultaneously

$\sum v_1w_2$ = the sum of ones in column v_1 when w_2 is one (or vice versa); hence, it is the number of observations in which the first level of A occurs with the second level of C, which is 2 in this example

The resulting X'X matrix is thus a table of frequencies or *incidences* of the occurrence of various treatment combinations and is often referred to as the *incidence matrix*.

The elements of the vector X'$\underline{y}$ are sums and partial sums. For example,

$\sum xy$ = the sum of y over all observations, usually denoted by $Y_{...}$, which is 168 in this example

$\sum v_1y$ = the sum of y when v_1 is one; hence, it is the sum of y for the first level of A, usually denoted by $Y_{1..}$, which is 56 in this example

$\sum w_2y$ = the sum of y when w_2 is one; hence, it is the sum of y for the second level of C, usually denoted by $Y_{.2.}$, which is 42 in this example

At this point, regression procedures call for the estimation of coefficients (μ, α_i, γ_j) by obtaining the solution to the normal equations, using X'X as the matrix of coefficients and X'$\underline{y}$ as the right-hand side. This is usually accomplished by obtaining the inverse of X'X, which is also used for various tests. Actually, this matrix is singular and will not invert; this is discussed in Section 7.3. For this case it is, however, possible to obtain directly the estimates and associated tests.

The first row of the matrix corresponds to the equation

$$24\mu + 8\alpha_1 + 8\alpha_2 + 8\alpha_3 + 6\gamma_1 + 6\gamma_2 + 6\gamma_3 + 6\gamma_4 = 168$$

However, the restrictions on the parameters [see equation (7.2)]

$$\sum \alpha_i = \sum \gamma_j = 0$$

cause elimination of the terms involving α and γ; hence, the equation becomes

$$24\mu = 168$$
$$\hat{\mu} = 7$$

In other words, μ is estimated by the grand mean, $\bar{y}_{...}$ obtained by the usual analysis. The second equation is

$$8\mu + 8\alpha_1 + 2\gamma_1 + 2\gamma_2 + 2\gamma_3 + 2\gamma_4 = 56$$

Again the restriction eliminates the γ coefficients, and substituting for $\hat{\mu}$ as previously estimated

$$8\hat{\alpha}_1 = 56 - (8)(7)$$
$$\hat{\alpha}_1 = 0$$

which was also obtained previously. In a similar manner,

$$\hat{\alpha}_2 = 2 \qquad \hat{\alpha}_3 = -2 \qquad \hat{\gamma}_1 = -2$$
$$\hat{\gamma}_2 = 0 \qquad \hat{\gamma}_3 = 2 \qquad \hat{\gamma}_4 = 0$$

which are the same as the usual estimates of the treatment effects. In terms of the regression problem, the vector of estimates is

$$\underline{\hat{\beta}}' = [7 \quad 0 \quad 2 \quad -2 \quad -2 \quad 0 \quad 2 \quad 0]$$

The sum of squares due to regression is

$$\begin{aligned} SSR &= \underline{\hat{\beta}}'\underline{g} \\ &= 7(168) + 0(56) + 2(72) + -2(40) + -2(30) + 0(42) \\ &\quad + 2(54) + 0(42) = 1288 \end{aligned}$$

Usually, individual terms of $\underline{\hat{\beta}}'\underline{g}$ have no meaning. However, in this case a division into portions provides some useful results:

1. Term 1:

$$\begin{aligned} \bar{y}_{...}Y_{...} &= \frac{Y_{...}}{acn} \\ &= \text{correction factor for mean} \\ &= \text{SS due to mean} \end{aligned}$$

2. Term 2 + 3 + 4:

$$\begin{aligned} &(\bar{y}_{1..} - \bar{y}_{...})Y_{1..} + (\bar{y}_{2..} - \bar{y}_{...})Y_{2..} + (\bar{y}_{3..} - \bar{y}_{...})Y_{3..} \\ &= \sum nc\bar{y}_{i..}(\bar{y}_{i..} - \bar{y}_{...}) = \sum nc(\bar{y}_{i..} - \bar{y}_{...})^2 \\ &= \text{SS due to the A treatment (by the usual methods)} \end{aligned}$$

3. Finally, the sum of terms 5 through 8 becomes the sum of squares due to the C treatment. Numerically, for the example

   Total SS $= \sum \sum y^2 = 1382$
   SS due to mean = 1176
   SS due to A = 64
   SS due to C = 48

By subtraction, the residual sums of squares is

$$1382 - 1176 - 64 - 48 = 94$$

which can easily be verified from the previous analysis in Table 7.2 as

$$\text{SS interaction} + \text{SS error} = 86 + 8 = 94$$

We may now ask: Why was it not necessary to invert the matrix; the matrix is *not* diagonal. The reason is simple: these data come from a *balanced* experiment, that is, each treatment combination occurs an equal number of times (twice); therefore the γ cancel out in the equations that are solved for the α, and vice versa. It is also for this reason that individual sets of terms of $\hat{\underline{\beta}}'\underline{g}$ could be used to obtain sums of squares. This is, however, not true in the case of analyses of unbalanced data, where each treatment combination is not replicated the same number of times. For this reason, such data *cannot* be analyzed in the usual manner; this is the subject of the next section.

7.3 TWO-FACTOR UNBALANCED EXPERIMENT

This case will also be illustrated with a small artificial example, this one with two levels of A and four levels of C and an unequal number of replications within each AC combination (cell). The data are given in Table 7.5.

The model for the analysis of these data is the same as in the previous section; the interaction is again to be obtained by subtraction. The error component of the model is estimated independently by a between-within-cell ANOVA,[†] which is given in Table 7.6. The seven degrees of freedom between cells sum of squares corresponds to the total effects of A, C, and (A × C); thus the statistically significant F ratio in this analysis indicates that one or more of the effects (α, γ, or $\alpha\gamma$) is not zero, but it does not indicate *which* of these contribute to this result.

The least-squares analysis is implemented by the introduction of dummy variables, resulting in a design matrix of "observed" ones and zeros; the sums of squares and cross products comprise the incidence matrix X'X and the right-hand side X'$\underline{y}$, which can be derived directly by the counting and summing procedures described in the previous section. These matrices are given in Table 7.7.

[†]Lack of balance (unequal replications) is permitted in a "one-way" analysis of variance.

TABLE 7.5 Two-Factor Experiment: Unequal Frequencies

A Treatment	C Treatment 1	2	3	4	Total
1	4	9	5	7	
	5	7		10	63
	6			10	
2	7	8	3	13	
	7	12	5	15	97
	9		7		
	11				
Total	49	36	20	55	160

TABLE 7.6 Between-Cell Analysis

Source	d.f.	SS	MS	F
Between cells	7	147	21.0	6.4688[a]
Within cells	12	39	3.25	
Total	19	186		

[a]Statistically significant at the 0.01 level.

The fact that for this example it is not possible to solve directly for the coefficients (effects) is readily apparent. The first equation, which provides the estimate for μ, is

$$20\mu + 9\alpha_1 + 11\alpha_2 + 7\gamma_1 + 4\gamma_2 + 4\gamma_3 + 5\gamma_4 = 160 = Y_{...}$$

However, the restriction

$$\sum \alpha_i = \sum \gamma_j = 0$$

does *not* eliminate any terms; thus, $\bar{y}_{...}$, the usual estimate of μ, is a function of all parameters. Specifically it can be seen that

TABLE 7.7 Incidence Matrix

Parameter	Variable	(X'X)							$X'\underline{y}$
		x	v_1	v_2	w_1	w_2	w_3	w_4	
μ	x	20	9	11	7	4	4	5	160
α_1	v_1	9	9		3	2	1	3	63
α_2	v_2	11		11	4	2	3	2	97
γ_1	w_1	7	3	4	7				49
γ_2	w_2	4	2	2		4			36
γ_3	w_3	4	1	3			4		20
γ_4	w_4	5	3	2				5	55

$$\bar{y}_{...} = \mu + \frac{1}{20}(9\alpha_1 + 11\alpha_2 + 7\gamma_1 + 4\gamma_2 + 4\gamma_3 + 5\gamma_4)$$

This expression may be simplified somewhat:

$$\bar{y}_{...} = \mu + \frac{1}{20}(2\alpha_2 + 3\gamma_1 + \gamma_4)$$

but it *cannot* be made purely a function of μ.

Further, for example, it can be seen that the usual estimate for γ_3 is

$$\bar{y}_{.3.} - \bar{y}_{...} = \mu + \gamma_3 + \frac{1}{4}(\alpha_1 + 3\alpha_2) - \mu - \frac{1}{20}(-2\alpha_1 + 3\gamma_1 + \gamma_4)$$

which may be simplified, but in any case this estimate of γ_3 depends on the actual values of the α_i as well as the other γ_j. Actually the estimate $(\bar{y}_{.3.} - \bar{y}_{...})$ is equivalent to the estimate of a total regression coefficient, that is, ignoring the effect of other factors. Therefore, it is necessary to estimate all terms *simultaneously*, which is accomplished by solving the *set* of normal equations in μ, α, and γ, whose coefficients and right-hand side are given in Table 7.7.

This is, however, not immediately possible due to linear dependencies among the equations. For example, the equation for μ is the sum of the equations for the α_i, as well as for the γ_j. Therefore, the equation for μ is *linearly dependent* on the equations for the α_i and γ_j. In other words, there are not really seven *different* equations in the seven unknown parameters. Actually, with these two linear dependencies there are only five linearly independent equations. This produces a singular coefficient matrix for which there is no unique inverse (Section 1.4).

Linear models whose estimation require normal equations with singular coefficient matrices are known as *models of less than full rank*. The number of parameters which can be uniquely estimated is limited by the rank of the matrix; in the present example, only five parameters can be estimated. The result is that there is not a unique solution; in fact, there exist an infinite number of solutions and associated estimates of parameters that satisfy the original set of seven equations.

Two different approaches to obtaining useful estimates of parameters for models of less than full rank have been developed. One method, which is very general but somewhat difficult to implement, uses generalized inverses and estimable functions of the parameters. The basic premise of this method is that certain linear functions of the parameters can be uniquely estimated. In the present example, the following five functions of parameters can be uniquely estimated:

1. μ
2. $\alpha_1 - \alpha_2$
3. $-3\gamma_1 - \gamma_2 + \gamma_3 + 3\gamma_4$
4. $\gamma_1 - \gamma_2 - \gamma_3 + \gamma_4$
5. $-\gamma_1 + 3\gamma_2 - 3\gamma_3 + \gamma_4$

Normal equations for obtaining estimates of these functions are, indeed, of full rank, and a unique set of estimates can be obtained. There are, however, an infinite number of sets of such functions; these are called *estimable functions* of the parameters. Procedures have been developed whereby a general solution of the normal equations is obtained, and mechanics are then provided whereby desired estimable functions are generated by the user and corresponding estimates and tests are obtained from the general solution. Further exposition of this methodology is beyond the scope of this book and will not be presented.

The other approach to obtaining a set of useful estimates employs linear restrictions on the set of parameters. This methodology works well for most applications and provides estimates and tests that are equivalent to those obtained by the estimable-functions procedure. However, in special cases, particularly when a number of treatment combinations have no observations (missing cells), the resulting parameter estimates may not be meaningful. The present example has no missing cells and exemplifies the most common application of linear model methodology.

A number of different sets of restrictions on the parameters may be employed to obtain estimates. For example, one method is to arbitrarily set one of the parameter estimates of each effect equal to zero. Most restrictions produce equivalent results; however, they must be used with caution. The most commonly used restrictions arise from the usual restrictions on the linear model [equation (7.2)]:

$$\sum \alpha_i = \sum \gamma_j = 0$$

These restrictions are restated as follows:

$\alpha_{last} = -\sum \alpha_i$, with the summation covering the α_i other than the last one

and

$\gamma_{last} = -\sum \gamma_j$, with the summation covering the γ_j other than the last one

or, in the example,

TABLE 7.8 Reduced Design Matrix

A	C	Reps.	x	v_1	w_1	w_2	w_3	y
1	1	1	1	1	1			4
		2	1	1	1			5
		3	1	1	1			6
	2	1	1	1	1	1		9
		2	1	1		1		7
	3	1	1	1			1	5
	4	1	1	1	-1	-1	-1	7
		2	1	1	-1	-1	-1	10
		3	1	1	-1	-1	-1	10
2	1	1	1	-1	1			7
		2	1	-1	1			7
		3	1	-1	1			9
		4	1	-1	1			11
	2	1	1	-1		1		8
		2	1	-1		1		12
	3	1	1	-1			1	3
		2	1	-1			1	5
		3	1	-1			1	7
	4	1	1	-1	-1	-1	-1	13
		2	1	-1	-1	-1	-1	15

TABLE 7.9 Reduced X'X and X'$\underline{y}$ Matrices

Vars.	x	v_1	w_1	w_2	w_3	y
x	20	-2	2	-1	-1	160
v_1	-2	20	-2	-1	-3	-34
w_1	2	-2	12	5	5	-6
w_2	-1	-1	5	9	5	-19
w_3	-1	-3	5	5	9	-35

$$\alpha_2 = -\alpha_1$$

$$\gamma_4 = -(\gamma_1 + \gamma_2 + \gamma_3)$$

Inserting these restrictions into the model reduces the number of parameters to five: μ, α_1, γ_1, γ_2, and γ_3.

These restrictions are incorporated into the analysis by a corresponding specification of the associated dummy variables. In general, for a factor having p levels, there are p - 1 corresponding dummy variables, say v_i, defined as follows:

For observations occurring in levels 1, 2, ..., p - 1,

$$v_i = \begin{cases} 1 & \text{if observation is in i-th level} \\ 0 & \text{otherwise} \end{cases}$$

For observations occurring in level p,

$$v_i = -1 \qquad (i = 1, 2, \ldots, p - 1)$$

The design or X matrix obtained by the application of these dummy variable definitions is often called the *reduced* matrix, since its order has been reduced to reflect the number of parameters to be estimated. Table 7.8 gives the reduced X matrix for the current example. This is used to obtain reduced X'X and X'$\underline{y}$ matrices in the usual manner; these are given in Table 7.9. Since these are not tables of frequencies and partial sums, they are obtained by calculating sums of squares and cross products.

The corresponding normal equations are solved (Table 7.10), resulting in the vector of estimates:

$$\hat{\alpha}_1 = -1.500766$$

$$\hat{\gamma}_1 = -1.048239$$

$$\hat{\gamma}_2 = 1.166156$$

$$\hat{\gamma}_3 = -3.584227$$

The restrictions are invoked to obtain

TABLE 7.10 Reduced Inverse (C matrix)

	μ	α_1	γ_1	γ_2	γ_3
μ	0.05333556	0.00593415	-0.01677354	0.00916443	0.01213150
α_1	0.00593415	0.05359877	0.00172281	-0.00593415	0.02086523
γ_1	-0.01677354	0.00172281	0.12416252	-0.04572645	-0.04486504
γ_2	0.00916443	-0.00593415	-0.04572645	0.17833557	-0.07463150
γ_3	0.01212150	0.02086523	-0.04486504	-0.07463150	0.18580111

$$\hat{\alpha}_2 = 1.500766$$

$$\hat{\gamma}_4 = 3.466310$$

The sums of squares needed for the analysis of variance for testing H_0: $\underline{\beta} = 0$ are obtained in the usual way, e.g.,

$$\text{SS regression} = \hat{\underline{\beta}}'\underline{g} = 1414.0213$$

which includes the SS due to the mean. The desired test is for the contribution of effects to the sum of squares adjusted for the mean; hence, the correction for the mean (Section 3.7, statement 7) $Y^2_{...}/\Sigma\, n = 160^2/20 = 1280$ is subtracted, thus

$$\text{SS}(\alpha \text{ and } \gamma) = 1414.0213 - 1280 = 134.0213$$

The between-cell sum of squares (Table 7.6) is the sum of squares due to α, γ, and $\alpha\gamma$; hence, the difference

$$\text{SS between cells} - \text{SS } (\alpha \text{ and } \gamma) = 147 - 134.0123 = 12.986$$

is due to interaction $\alpha\gamma$ and is denoted SSAC.[†]

[†]Note that SSAC is "adjusted," since it is calculated by direct subtraction (see Chapter 3). However, SSA and SSC are adjusted for *only* the effects *in* the model; in this case they are not adjusted for the AC interaction. See Section 7.5.

The complete analysis also requires the (adjusted) sums of squares attributable to the α and γ effects, respectively. They are obtained by using the formula for the sums of squares due to a subset of coefficients [formula (3.24)]. The resulting sums of squares are divided by the number of estimated parameters in the corresponding subset to obtain a mean square, which is in turn divided by the appropriate error mean square to obtain the F ratio for the test. In the example,†

$$\text{SS due to } \alpha = -1.5008(0.0536)^{-1}(-1.5008) = \frac{(-1.5008)^2}{0.0536} = 42.0224$$

and

SS due to γ

$$= [-1.0483 \quad 1.1662 \quad -3.584]\begin{bmatrix} 0.1242 & -0.0457 & -0.0449 \\ -0.0457 & 0.1783 & 0.0746 \\ -0.0449 & 0.0746 & 0.1858 \end{bmatrix}^{-1}\begin{bmatrix} -1.0483 \\ 1.1662 \\ -3.5842 \end{bmatrix}$$

$$= 117.6578$$

Note that, as in ordinary multiple regression, the *sum* of the two adjusted sums of squares has no real meaning. The results can be summarized in Table 7.11.

A further use of the inverse, as in regression, is to obtain confidence intervals on the individual coefficients. It is also possible to compute a t test for each effect. The confidence interval for any coefficient β_i is

$$\hat{\beta}_i \pm t\sqrt{c_{ii}s^2}$$

The corresponding 95% confidence interval for α_1 is

$$-1.5008 \pm 2.179\sqrt{(0.053599)(3.2500)}$$

---

†Only four decimals are presented here; the actual computations carried eight digits.

TABLE 7.11 Analysis of Variance

Source	d.f.	SS	MS	F
Due to A and C	4	134.0213	--	--
A,A, adj. for C	1	42.0224	42.0224	12.9300
C, adj. for A	3	117.6578	39.2193	12.0674
A × C, adj. for A and C	3	12.9786	4.3262	1.3311
Error	12	39.0000	3.2500	

and the resulting limits of the interval are

-2.4103 to -0.5913

Likewise, the 95% confidence limits on the other effects are

γ_1: -2.4325 to 0.3359

γ_2: -0.4925 to 2.8251

γ_3: -5.2775 to -1.8909

The intervals for α_2 and γ_4 are a little more difficult to compute, requiring the use of the restricting relationships and corresponding formulas for variances of linear functions which are discussed in Section 7.4. The results of these calculations are given in Table 7.12.

The major treatment differences occur at levels 3 and 4 of C. However, restrictions on the use of individual tests (e.g., the least significant difference) apply here as in the balanced case. Approximations for range tests are available but beyond the scope of this book. Contrasts, which are actually estimable functions, may be constructed, but will usually not be orthogonal; hence, they must be computed by using multiple regression, and the corresponding sums of squares will not partition the treatment sum of squares.

TABLE 7.12 Final Summary of Results

Effects	Estimates	95% Limits	
		Lower	Upper
Mean	7.8338	6.8971	8.7706
A	Statistically significant		
α_1	-1.5008	-2.4398	-0.5617
α_2	1.5008	0.5617	2.4398
C	Statistically significant		
γ_1	-1.0483	-2.4775	0.3810
γ_2	1.1662	-0.5468	2.8791
γ_3	-3.5842	-5.3326	-1.8358
γ_4	3.4663	1.8546	5.0079

7.4 VARIANCES OF TREATMENT DIFFERENCES AND COMBINATIONS

It is possible to compute variances, hence t tests and/or confidence intervals for treatment means, treatment differences, or other linear combinations of treatment effects. For these computations, it is necessary to use the formula for the variances of linear functions of random variables as discussed in Section 3.10 [formulas (3.34a) and (3.34b)].

For example, a treatment mean, e.g., the mean response of all observations in the i-th level of a treatment, say μ_i, is estimated by

$$\hat{\mu}_i = \hat{\mu} + \hat{\tau}_i$$

By formula (3.34a), since $a_1 = a_2 = 1$, the variance of this quantity is

$$V(\hat{\mu}_i) = V(\hat{\mu}) + V(\hat{\tau}_i) + 2\ \mathrm{Cov}(\hat{\mu}, \hat{\tau}_i)$$

which in turn can be written (see Section 3.8)

$$V(\hat{\mu}_i) = (c_{\mu\mu} + c_{\tau_i\tau_i} + 2c_{\mu\tau_i})\sigma^2$$

where the subscripts "locate" elements in the C matrix with respect to the parameters (see Table 7.10); the error mean square is used to estimate σ^2.

In a similar manner, the estimated difference between two treatment means is

$$\hat{\delta}_{ij} = \hat{\tau}_i - \hat{\tau}_j$$

and its variance is

$$V(\hat{\delta}_{ij}) = (c_{\tau_i\tau_i} + c_{\tau_j\tau_j} - 2c_{\tau_i\tau_j})\sigma^2$$

The preceding principle is also applied to obtain estimates of those coefficients which were deleted in the reduction process. If, for example, there are four treatment levels,

$$\hat{\tau}_4 = -\sum_{i=1}^{3} \hat{\tau}_i$$

All $a_i = -1$; hence,

$$V(\hat{\tau}_4) = (\sum_{i=1}^{3} \sum_{j=1}^{3} -1(-1)c_{\tau_i\tau_i})\sigma^2$$

Note that the coefficient of σ^2 is the sum of all the elements of the C matrix corresponding to the first three levels of the τ treatment. In the example of the previous section,

$$\begin{aligned} \text{Var}(\hat{\gamma}_4) &= 3.25(0.1242 - 0.0457 - \cdots - 0.0746 + 0.1858) \\ &= 0.5130 \end{aligned}$$

This method was used to obtain the confidence intervals for α_2 and γ_4 in Table 7.12.

The above procedure can also be used to obtain covariances involving "reduced" terms. The covariance of the reduced term, say τ_p, and another term of the same set, τ_i, is

$$\text{Cov}(\hat{\tau}_i, \hat{\tau}_p) = \text{Cov}[(\hat{\tau}_i), (-\sum_j \hat{\tau}_j)]$$
$$= -\sum_j (\text{Cov } \hat{\tau}_i, \hat{\tau}_j)$$

where the summation is again over the n - 1 "nonreduced" terms of that effect. This covariance is then estimated by

$$\text{Cov}(\hat{\tau}_i, \hat{\tau}_p) = -\sum_j c_{\tau_i\tau_j} \sigma^2$$

Similarly, the covariance of the reduced term and a nonreduced term of another set, say γ_i,

$$\text{Cov}(\hat{\gamma}_i, \hat{\tau}_p) = \text{Cov}[(\hat{\gamma}_i), (-\Sigma \hat{\tau}_j)]$$
$$= -\sum_j \text{Cov}(\hat{\gamma}_i, \hat{\tau}_j)$$

where the summation is again over the unreduced terms of both τ and γ effects. As above, this is estimated by

$$\text{Cov}(\hat{\gamma}_i, \hat{\tau}_p) = -(\sum c_{\gamma_i\tau_j})\sigma^2$$

7.5 UNBALANCED DATA, MULTIFACTOR EXPERIMENTS

For the analysis of multifactor experimental data, straightforward extensions of the above are used. These procedures will not, however, obtain estimates of *individual* interactions. For example, if in a 4 × 5 × 7 factorial array, the model is specified to be

$$y_{ijkl} = \mu + \alpha_i + \gamma_j + \delta_k + \varepsilon_{ijkl}$$

the ANOVA table will be of the form:

Source	d.f.
Among cells	139
A	3
B	4
C	6
All interactions	126
Within	$\sum\sum\sum n - 140$

In this case, the difference between the *among-cell* sums of squares with 139 degrees of freedom and the total sum of squares due to A, B, and C (main effects only) will be a combined sum of squares due to all interactions (AB, A, BC, ABC), or more accurately, due to the lack of fit of the model as used for the basis for the incidence matrix. The main effect estimates and sums of squares will not be adjusted for interactions.

The F ratio of the mean square for all interactions divided by the *within* mean square indicates whether interactions may exist, although it is quite possible that one or more individual interactions exist while the overall test is nonsignificant.

It is possible to specifically include interactions in the model and obtain tests for individual interactions as well as estimates of individual interaction effects. This is accomplished by the addition of more dummy variables. For example, to estimate the AB interaction in the above example, the following additional terms would be used:

$$y = \mu + \cdots + \sum_i \sum_j z_{ij} (\alpha\beta)_{ij}$$

where $z_{ij} = 1$ for any observation of the i-th A-factor and j-th B-factor levels, and is 0 for all others. In the above case, this would add 4(5) = 20 such additional terms. The resulting matrix contains additional linear dependencies due to the restrictions on the interaction terms:

$$\sum_i \alpha\beta_{ij} = 0 \qquad (j = 1, 2, \ldots, 5)$$

$$\sum_j \alpha\beta_{ij} = 0 \qquad (i = 1, 2, \ldots, 4)$$

These are nine restrictions, but one is defined by the others, and hence there are eight linear dependencies, leaving 12 parameters corresponding to the 12 degrees of freedom for this interaction.

The appropriate reduced interaction effects can be produced directly by creating, in the reduced X matrix, dummy variables from all possible pairwise cross products of the corresponding main effect dummy variables.

The analysis of variance based on this addition to the model would be

Source	d.f.
A	3
B	4
C	6
AB	12
Other interactions	114
Within	$\sum\sum\sum n - 140$

The inclusion of interactions can easily produce very large X'X matrices which may be expensive to invert even with high-speed computers. Thus, care should be taken to include only such interactions which are of real interest. However, effects estimates are only adjusted for those effects actually included in the model. Further, analysis procedures and results are complicated by missing treatment combinations (see comments in Section 7.3).

It is quite obvious that the use of the methodology described in this chapter is computationally tedious and must largely be implemented by the use of high-speed computers. Fortunately, computer programs to accomplish this type of analysis are available in all major statistical software packages. Users of such packages should, however, be cautioned to read carefully the documentation of the program to be used, as not all of these routines necessarily use the same procedure for obtaining unique coefficient estimates. In fact, at least one computer program for this type of analysis (the GLM procedure of SAS-76) uses the methodology of estimable functions instead of imposing restrictions.

7.6 PROBLEMS

1. (a) This is a 3 × 3 factorial with unequal replications. Analyze by least squares and obtain the interaction by subtraction. (Obtain inverse and complete solution.)

			C	
		1	2	3
	1	17.6	27.3	16.7
		16.6	20.9	
		16.3		
	2	31.4	23.5	25.5
A		32.5	25.6	22.4
			23.7	23.6
			22.1	
	3	23.0	27.3	29.6
		31.8	34.2	34.6
		29.4	28.5	36.0
		30.9		27.8
				30.2

(b) Analyze as if balanced, using formulas of the type

$$SSA = \sum \frac{Y_{i..}^2}{n_{i.}} - \frac{Y_{...}^2}{n_{..}}$$

and compare results.

2. *Complementation* design. Assume that two chemical ingredients (A and C) are mixed in varying proportions in a neutral solution (given as percentages of the total mixture) and that the composition of the mixture has an effect on some reaction Y. Since the ingredients cannot total more than 100%, there is a restriction that $A + C \leq 100$. Suppose the following results have been obtained:

	C				
A	0	25	50	75	100
0	14	21	29	45	62
25	10	20	21	30	
50	5	18	18		
75	2	15			
100	5				

Set up for solution by least squares to estimate effects of A and C, and proceed with obtaining a solution *without* inverting the matrix.

3. An incomplete blocks experiment is used in situations where block sizes are restricted to be smaller than the desired number of treatments. A balanced incomplete blocks experiment is constructed such that all pairs of treatments occur equally often within blocks. Consider a balanced incomplete blocks experiment with

   t = 4 treatments
   k = 2 units per block
   b = 6 blocks

   as follows:

Blocks	1	2	3	4	5	6
Treatments	1	3	1	2	1	2
	2	4	3	4	4	3

Obtain normal equations (unreduced) for estimating the τ_i (treatment effects) and β_j (block effects), and obtain an expression for $\hat{\tau}_1$ without inverting the matrix. (Hint: Use equation for $\hat{\tau}_1$ *and* equations corresponding to unwanted parameters in that equation. Note that $\bar{y}_{..}$ is an estimate of μ.)

4. Least squares for missing plot. This problem illustrates how the least-squares method is equivalent to a well-known procedure for handling the case of a single missing observation in a two-factor experiment with no replications.

(a) Given the following 2 × 3 factorial with the second A- and first B-factor observation (y_{21}) missing.

A × B	1	2	3	Totals
1	4	5	6	15
2	(M)	9	6	15
Totals	4	14	12	30

Set up and solve by least squares for A and B main effects; complete with ANOVA table. Since there are no replications the A × B interaction must be used as error.

(b) The standard procedure for this situation is given by Ostle and Mensing (1975, Section 11.12, pp. 408-409). First estimate the missing value by the formula

$$M_{ij} = \frac{aY_{i.} + bY_{.j} - Y_{..}}{(a - 1)(b - 1)}$$

where

M_{ij} = missing value

a, b = number of levels of the A and B treatments

Next, perform the analysis of variance with the completed data. Subtract *one* degree of freedom from SSE (which would normally be 2 for this example). Finally, perform an adjustment on the sums of squares; for A, this adjustment is to obtain

$$\text{SSA, adj} = \text{SSA} - Z_A$$

where

$$Z_A = \frac{Y_{.j} - (a - 1)M_{ij}^2}{a(a - 1)}$$

The adjustment factor for B is, similarly,

$$SSB, adj = SSB - Z_B$$

where

$$Z_B = \frac{Y_{i.} - (b - 1)M_{ij}^2}{b(b - 1)}$$

Complete this analysis and compare with part (a).

5. Five objects are weighed in sets of three and the total weight of a set is recorded: y_{ijk} is the total weight of items i, j, and k. We weigh all 10 combinations of three as follows:

y(123)	5
y(124)	7
y(125)	8
y(134)	9
y(135)	9
y(145)	12
y(234)	8
y(235)	8
y(245)	11
y(345)	11

Set up a model and *estimate* the weights of the *individual items*; assume there is a random weighing error. You will *not* need to invert the matrix and you will *not* be required to furnish anything except the estimates of the weights. (Hint: It is easier if you do not correct for the mean.) Assume you are told the weighings are *exact*. If they, indeed, are exact, would you use a different estimating procedure? If so, specify the procedure. How can this assumption be tested?

Chapter 8

REGRESSION WITH GROUPED DATA; COVARIANCE

8.1 INTRODUCTION

The methodology presented in this chapter is based on models which use *both dummy* and continuously measured independent variables. In other words, it is a combination of regression and the analysis of variance and is used not only to establish regression relationships but also to differentiate the effects of treatments or groups.

For example, the auction market data in Chapter 2 were collected in 1967, and equivalent data are available for 1969; it is of interest to determine whether the same relationship between cost and sales volume holds for both years. As another example, it is known that knowledge gained is related to aptitude; this is established by a regression model. It is of interest to determine whether this relationship is the same across different teaching treatments. Possibly some teaching methods are better for low-aptitude students, while others are better for high-aptitude students.

This fusion of regression and the analysis of variance provides for a variety of models and associated hypothesis tests and estimates. Three alternate models are commonly used:

1. The regression relationship is the same among groups or treatments. In other words, one regression model, specified by a single β-vector, adequately describes the entire set of data.
2. The regression relationship is the same for all groups, except there is a difference in the *level* of response among

groups. In other words, a single β-vector describes the relationship, but there may be different intercepts (β_0-values). In the auction market example, this situation would occur if the costs of selling the different classes of livestock remained the same over both years but higher fixed costs were experienced in the latter year.

3. One set of regression coefficients is not adequate to describe the data. In other words, different regression vectors, including both β_0 and the other partial regression coefficients, may be needed for the different groups. In the auction market example this would correspond to the possibility that, due to inflation or marketing conditions, the costs of selling the different classes of animals changed between the two years.

A relatively straightforward procedure for testing which of these models is most appropriate is provided in Section 8.2. Model 2 is usually called the *analysis of covariance*, which is thoroughly discussed in most basic statistics texts; it is briefly discussed in Section 8.4 and is related to the procedures of Section 8.2. The three models may also be considered as special cases of a single general linear model and the appropriate model chosen by tests regarding subsets of the coefficients of the model; this is presented in Section 8.5.

An additional model may be used to establish a regression relationship among the group or treatment means of the dependent and the independent variables. Further, if such a relationship exists, there is a test to determine whether the regression among means is related to the regression among the individual observations, if model 1 or 2 has been accepted. The appropriate tests associated with the models relating group means are given in Section 8.3.

It is quite obvious that most practical applications of these models require the use of high-speed computers. Discussions of the computational aspects of implementing these models are given in Section 8.6.

8.2 MODELS INVOLVING INDIVIDUAL OBSERVATIONS

Consider data divided into p groups with n_i observations in the i-th group. The k-th observation in the i-th group consists of the dependent variable y_{ik} and m associated independent variables x_{jik}; j = 1, 2, ..., m; i = 1, 2, ..., p; k = 1, 2, ..., n_i.

The discussion of the models is illustrated with a very simple example of a one-variable linear regression in p = 3 groups of n_i = 5. The simplicity of the example allows graphic as well as numeric illustration. The data are presented in Table 8.1. A cursory inspection would seem to indicate that model 2 may be most appropriate and that there is no regression among means.

TABLE 8.1 Data for Group Regression

	Group					
	1		2		3	
	x	y	x	y	x	y
	1	2	3	4	5	1
	2	2	4	5	6	3
	3	3	5	5	7	4
	4	4	6	5	8	5
	5	4	7	6	9	7
Sums	15	15	25	25	35	20
Means	3	3	5	5	7	4
Sum of squares	55	49	135	127	255	100
Sum of crossproducts	51		129		154	

Total (all groups)

Sum (x) = 75	Mean (x) = 5	Sum of squares (x) = 445
Sum (y) = 60	Mean (y) = 4	Sum of squares (y) = 276

Sum of crossproducts = 334

The procedure for determining the most appropriate model is to first estimate the parameters of the three models as follows:

1. Model 1, which contains m + 1 parameters, that is, the elements of one β-vector and one β_0-value
2. Model 2, which contains m + p parameters, that is, the elements of one β-vector and p β_0-values
3. Model 3, which contains p(m + 1) parameters, that is, p sets of m-element β-vectors and p β_0-values

The appropriate partitioning of sum of squares is performed for each of the models. A step-down analysis of variance, starting with model 3, is used to determine whether the increase in the error sum of squares allows the use of the fewer parameters of models 2 or 1, respectively.

8.2.1 Model 1: One Regression

In this model the grouping of the data is ignored; in other words, one regression is performed on the entire data set of Σn_i observations. Estimates of the coefficients and the partitioning of the sums of squares proceed normally, producing results of the form given in Table 8.2.

In the example,

$$\hat{\beta}_1 = \frac{34}{70} = 0.4857$$

and

TABLE 8.2 Model 1: Partitioning of Sums of Squares

Source	d.f.	SS
Total (1)	$(\sum n_i) - 1$	SST(A)
Regression (1)	m	SSR(A)
Residual (1)	$(\sum n_i) - m - 1$	SSE(A)

TABLE 8.3 Summary Statistics, Model 1

Source	d.f.	SS
Total (1)	14	36
Regression (1)	1	16.5143
Residual (1)	13	19.4857

$$\hat{\beta}_0 = 4 - 0.4857(5) = 1.5715$$

The partitioning of the sum of squares is given in Table 8.3.

8.2.2 Model 2: Parallel Lines or Planes

This model provides for a common set of partial regression coefficients, $\underline{\beta}_w$, but allows for different intercepts (β_{0iw}, i = 1, 2, ..., p) for the groups; the subscript w is used to differentiate the parameters from those of model 1. Since intercepts are functions of means, this model essentially describes the data as having a common set of regression coefficients, if the data are adjusted or corrected for differences among group means. This is accomplished by first adjusting or correcting the sums of squares and cross products of *each* group for its respective mean and then summing these corrected sums of squares and products to obtain the elements of a single set of matrices, $X'X_w$ and $X'\underline{y}_w$. In other words, $X'X_w$ and $X'\underline{y}_w$ are the sums of the corrected $X'X$ and $X'\underline{y}$ matrices for each group.

The diagonal elements of $X'X_w$ are the familiar *pooled-within-group sums of squares* associated with the analysis of variance of the x_i, and the off-diagonal elements are equivalently the *pooled-within-group sums of cross products*. Thus, the partial regression coefficients in model 2 are the coefficients of a *pooled-within-group regression*. The β_0 values are calculated for each group separately, by the formula

$$\hat{\beta}_{oiw} = \bar{y}_{i.} - \hat{\beta}_{1w}\bar{x}_{1i.} - \cdots - \hat{\beta}_{mw}\bar{x}_{mi.} \qquad (i = 1, 2, \ldots, p)$$

The model contains m + p parameters: one β-vector with m elements and p intercepts.

The partitioning of sums of squares (Table 8.4) is performed using $X'X_w$ and $X'\underline{y}_w$; the total sum of squares is the pooled-within sum of squares for the dependent variable with $(\Sigma\, n_i) - p$ degrees of freedom.

In the example

$$X'X_w = 10 + 10 + 10 = 30$$

$$X'\underline{y}_w = 24$$

$$SST = 26$$

hence,

$$\hat{\beta}_{1w} = (X'X_w)^{-1}\, X'\underline{y}_w$$

$$= \frac{24}{30} = 0.8$$

$$\hat{\beta}_{01w} = \bar{y}_{1.} - \hat{\beta}_{1w}\bar{x}_{1.} = 3.0 - 0.8(3.0) = 0.6$$

$$\hat{\beta}_{02w} = \bar{y}_{2.} - \hat{\beta}_{1w}\bar{x}_{2.} = 5.0 - 0.8(5.0) = 1.0$$

$$\hat{\beta}_{03w} = \bar{y}_{3.} - \hat{\beta}_{1w}\bar{x}_{3.} = 4.0 - 0.8(7.0) = -1.6$$

$$SSR = 0.8(24) = 19.2 \qquad d.f. = 1$$

and

$$SSE = 6.8 \qquad d.f. = 11$$

The partitioning of sums of squares is summarized in Table 8.5.

TABLE 8.4 Model 2B: Partitioning of Sums of Squares

Source	d.f.	SS
Total (2)	$(\sum n_i) - p$	SSW
Regression (2)	m	SSR(B)
Residual (2)	$(\sum n_i) - m - p$	SSE(B)

TABLE 8.5 Summary Statistics, Model 2B

Source	d.f.	SS
Total (2)	12	26
Regression (2)	1	19.2
Residual (2)	11	6.8

8.2.3 Model 3: All Different Regressions

This is the most complex of the three models, since a total of $p(m + 1)$ coefficients and intercepts are estimated. Conceptually, however, it is quite simple, in that each group is treated as an independent sample and the regression analysis is performed separately for each group. Thus there is, for each group, the partitioning of the sums of squares: SST_i, SSR_i, and SSE_i, with $n_i - 1$, m, and $n_i - m - 1$ degrees of freedom, respectively; $i = 1, 2, \ldots, p$. The overall summary of the adequacy of this model is obtained by summing the partitions over the p groups; the general form of the resulting partitioning is given in Table 8.6.

In the example, the statistics of the individual group regression estimates are given in Table 8.7. The sums of the relevant sums of squares provide results given in Table 8.8.

TABLE 8.6 Model 3: Partitioning of Sums of Squares

Source	d.f.	SS
Total (3)	$(\sum n_i) - p$	$\sum_i SST_i = SST(C)$
Regression (3)	mp	$\sum_i SSR_i = SSR(C)$
Residual (3)	$\sum n_i - p(m + 1)$	$\sum_i SSE_i = SSE(C)$

TABLE 8.7 Example Computations of Model 3

Computation or statistic	Group 1	Group 2	Group 3
$X'X = \sum x^2 - (\sum x)^2/n$	10	10	10
$X'\underline{y} = \sum xy - \sum x \sum y/n$	6	4	14
$SST = \sum y^2 - (\sum y)^2/n$	4	2	20
$\hat{\underline{\beta}} = (X'X)^{-1}X'\underline{y}$	0.6	0.4	1.4
$\hat{\beta}_0$	1.2	3.0	-5.8
SSR	3.6	1.6	19.6
SSE	0.4	0.4	0.4
$d.f._e$	3	3	3

TABLE 8.8 Summary Statistics, Model 3

Source	d.f.	SS
Total (3)	12	26
Regression (3)	3	24.8
Residual (3)	9	1.2

8.2.4 Step-down Analysis

As previously indicated, the three models have been presented in increasing order of complexity as evidenced by the number of parameters estimated. The choice of the most appropriate model is made by a stepdown analysis which indicates whether decreasing complexity (fewer parameters) significantly increases the residual mean square.

The simplest model (model 1) estimates m + 1 parameters, and the residual sum of squares has $\Sigma n_i - m - 1$ degrees of freedom. Model 2 estimates m + p parameters, and the residual sum of squares has $\Sigma n_i - m - p$ degrees of freedom, or p - 1 *fewer* than the residual from model 1. Thus, any reduction in SSE from model 1 to 2 is the *reduction in residual (or addition to regression) sum of squares due to* p - 1 *additional intercepts*, with p - 1 degrees of freedom.

Model 3 has p(m + 1) parameters, hence Σn_i - p(m - 1) degrees of freedom for the residual sum of squares. The difference between model 2 and model 3 is then defined as the *reduction in residual (or addition to regression) sum of squares due to* m(p - 1) *additional coefficients with* m(p - 1) *degrees of freedom.* This information is summarized in a step-down analysis of variance table outlined in Table 8.9.

The entries in the table indicate how the sums of squares are obtained from entries in Tables 8.2, 8.4, and 8.6; the mean squares are calculated in the usual manner. The step-down approach starts with the most complex model and steps down to successively lower order models until further simplifications are statistically nonsignificant. In other words, model 3 is required unless the reduction to model 2 is nonsignificant (using F_3), model 2 is used unless reduction to 1 is nonsignificant (using F_2), and a "no-regression" model is accepted if F_1 is nonsignificant. Note that *any* combinations of significance are possible.

Table 8.10 gives the summary analysis of variance for the example. It is evident that the most complex model, that is, the model which requires individual regressions for each group, is appropriate. This contradicts the initial opinion obtained by the "cursory inspection"; a plot of the data will show that group 3 observations show a much steeper slope, which prevents acceptance of model 2. It is now in order to present the estimates of the three separate regressions, using all appropriate tests of significance. Such a summary is presented in Table 8.11; obviously, other formats for this presentation are available.

TABLE 8.9 Group Data Regression--Summary Analysis

Source of variation	d.f.	SS: source	MS	
Total	$\sum n - 1$	SST(1):Table 8.2		
One regression	m	SSR(1):Table 8.2	MSR(1)	F_1 = MSR(1)/MSE(1)
Residual (1)	$\sum n - m - 1$	SSE(1):Table 8.2	MSE(1)	
Additional intercepts	p - 1	(SSE(1))-(SSE(2))	MSR(2)	F_2 = MSR(2)/MSE(2)
Residual (2)	$\sum n - m - p$	SSE(2):Table 8.4	MSE(2)	
Additional coefficients	m(p - 1)	(SSE(2))-(SSE(3))	MSR(3)	F_3 = MSR(3)/MSE(3)
Residual (3)	$\sum n - p(m + 1)$	SSE(3):Table 8.6	MSE(3)	

TABLE 8.10 Example, Summary Analysis of Variance

Source	d.f.	SS	MS	F
Total	14	36		
One regression	1	16.5143	16.5143	F_1 = 11.0176
Residual (1)	13	19.4857	1.4989	
Additional intercepts	2	12.6857	6.3428	F_2 = 10.2601
Residual (2)	11	6.8000	0.6182	
Additional coefficients	2	5.6000	2.8000	F_3 = 21.0053
Residual (3)	9	1.2000	0.1333	

TABLE 8.11 Summary of Individual Regression Estimates

	Group		
	1	2	3
$\hat{\beta}_0$	1.2	3.0	-5.8
$\hat{\beta}_1$	0.6	0.4	1.4
F	27.000	12.000	147.000
MSE	0.133	0.133	0.133
d.f.	3	3	3

8.3 MODEL INVOLVING GROUP MEANS

In some examples of regressions with grouped data, particularly when there are many groups, it is of interest to see if there exists a regression among the group *means*. Usually, interest in this relationship is limited to cases where common regression coefficients have been found to adequately describe the data (model 1 or 2). Further, if a regression among group means is found to exist, it is of interest to see if this relationship is similar to the within-group regression.

8.3.1 Regression among Group Means

In this model, a regression relationship is estimated by using group means as the data and obtaining a regression sum of squares and an F test with an appropriate error term.

Specifically, the $X'X_b$ and $X'\underline{y}_b$ matrices are constructed with elements consisting of the "between-group" sums of squares and cross products: the diagonal elements of X'X are the between-groups sums of squares as calculated for the corresponding analysis of variance. The off-diagonal elements are calculated in an equivalent manner.

Thus $X'X_b$ can be obtained by subtracting $X'X_w$ (model 2) from X'X (model 1). The above matrices are then used to obtain $\hat{\beta}_b$ and the partitioning of the sum of squares due to the regression in the usual manner, with the *total* sum of squares being the between-group sum of squares for the dependent variable (with k - 1 degrees of freedom). The residual in this case (with k - m - 1 degrees of freedom) is a measure of lack of fit; this can be tested against the appropriate error [residual (1) or (2), Table 8.9] as determined by that analysis. Table 8.12 shows this analysis, assuming that a single regression (model 1) holds. A significant F = MS (regression)/MS (error) would indicate that a regression exists; a significant F = MS (lack of fit)/(MS error) would indicate that the linear regression, even if significant, is not "good enough." These models are illustrated with the same example, even though this example did not indicate acceptance of models 1 or 2:

$$X'X_b = 415 - 375 = 40$$

$$X'y_b = 310 - 300 = 10$$

$$SST_b = 250 - 240 = 10$$

Hence,

$$\hat{\beta}_{1b} = \frac{10}{40} = 0.25$$

$$\hat{\beta}_{0b} = \bar{y}_{..} - \beta_{1b}\bar{x}_{..} = 4 - 0.25(5) = 2.75$$

$$SSR_b = 0.25(10) = 2.5$$

$$SSE_b = 10 - 2.5 = 7.5$$

$$SSE(3) = 1.2$$

(since model 3 was accepted; see Table 8.10).

The analysis of variance is given in Table 8.13. From this analysis, it can be seen that the regression among means is not (as expected) statistically significant.

TABLE 8.12 Regression among Means: Partitioning of Sums of Squares

Source	d.f.	SS
Total	$k - 1$	SST_b
Regression	m	SSR_b
Lack of fit	$k - m - 1$	SSE_b
Error (A)	$\Sigma n_i - m - 1$	SSE(A)

TABLE 8.13 Example, Regression among Means

Source	d.f.	SS	MS	F
Total	2	10		
Regression	1	2.5	2.5	<1
Lack of fit	1	7.5	7.5	56.251
Error (C)	9	1.2	0.1333	

8.3.2 Regressions among Means Equal to Regressions within Groups

As previously indicated this model is appropriate only if models 1 or 2 are found to be appropriate. Basically it is desired to test the hypothesis

$$H_0: \underline{\beta}_w = \underline{\beta}_b$$

which is a multivariate test of the equivalence of two vectors. Without going into detail, the test statistic is [see, e.g., Morrison (1976, Section 4.2)]

$$F = \frac{(\hat{\underline{\beta}}_w - \hat{\underline{\beta}}_b)'[(X'X_w)^{-1} + (X'X_b)^{-1}](\hat{\underline{\beta}}_w - \hat{\underline{\beta}}_b)}{\text{appropriate error MS}}$$

where $X'X_w$ and $X'X_b$ are the X'X matrices used to estimate $\hat{\underline{\beta}}_b$ and $\hat{\underline{\beta}}_w$, respectively. The appropriate error term depends on the choice of models 1 or 2. The degrees of freedom are m for the numerator and the degrees of freedom previously obtained for the error term in the denominator.

8.4 ANALYSIS OF COVARIANCE

The analysis of covariance consists essentially of the addition of one or more continuous independent variables to an analysis of variance model. For more extended details of covariance analysis, see, e.g., Ostle and Mensing (1975, Chapter 13).

For example, the model for a one-variable covariance analysis of a one-way classification is usually written

$$y_{ij} = \mu + \alpha_i + \beta(x_{ij} - \bar{x}_i) + \varepsilon_{ij}$$

where the α_i are parameters associated with differences among group mean and β is a regression parameter associated with the independent variable x, which is often called a *covariate* or *concomitant variable*.

It can be shown that this is equivalent to the model

$$y_{ij} = \mu + \alpha_i + \beta x_{ij} + \varepsilon_{ij}$$

which can readily be seen to be model 2 presented in Section 8.2. The usual tests in a covariance analysis are

1. H_0: $\beta = 0$
2. H_0: $\alpha_i = 0$

However, the *primary* interest is usually focused on item 2, the test for differences among means, adjusted for the effects of the covariate. A standard covariance analysis of the example of Section 8.2 is given in the usual format in Table 8.14. A comparison of these results with those of Table 8.10 shows that the test for groups (adj.) is equivalent to the test for differences among intercepts. The test of significance for the regression coefficient β is

$$F = \frac{SSR/m}{SSE/(\sum n_i - m - k)}$$

which for the example is

$$F = \frac{19.2}{6.8/11} = 31.070$$

TABLE 8.14 Covariance Analysis: Example

Source	d.f.	xx	xy	yy	d.f.	SS	MS	F
Groups	2	40	10	10				
Error	12	30	24	26	11	6.8	0.618	
G + E	14	70	34	36	13	19.486		
Groups, adj.					2	12.686	6.343	10.260[a]

[a]Significant at 0.01 level.

with (1, 11) degrees of freedom; this is identical to the result presented in Table 8.5.

8.5 THE GENERAL LINEAR MODEL

All of the above models which involve individual observations can be analyzed by specific tests on one general model. Such a model, if appropriately implemented, may be used for more specific tests, such as, for example, which coefficient(s) contributes to the differences in regression models among groups.

The general model is discussed in two stages: (1) analysis of covariance and (2) more general models.

8.5.1 Analysis of Covariance

The general model for covariance in a completely randomized two-factor experiment, for example, is

$$y_{ijk} = \mu + \alpha_i + \gamma_j + (\alpha\gamma)_{ij} + \beta x_{ijk} + \varepsilon_{ijk}$$

where α_i, γ_j, and $(\alpha\gamma)_{ij}$ are the parameters of the factorial experiment and βx_{ijk} is the regression or covariance portion.

The analysis of this is a relatively straightforward extension of the procedures discussed in Chapter 7. Table 8.15 provides a schematic of a typical normal equations matrix. Linear dependencies are eliminated from the dummy variable portion as discussed in Chapter 7.

TABLE 8.15 Normal Equations Matrix: Covariance Analysis

	X'X		
	Discrete factors (μ, α, β, γ, etc.)	Covariates	$X'\underline{y}$
Discrete	Incidence matrix of frequencies	Partial sums of the x variables	Partial sums of y
Covariates	Partial sums of the x variables	Uncorrected SS and SP of the x variables	Uncorrected SP of y with the x variables

The procedure is illustrated with the example of the Section 8.2. The covariance model is

$$y_{ij} = \mu + \tau_i + \beta x_{ij} + \varepsilon_{ij}$$

where the τ_i are the group effects. The reduced normal equations for this example are given in Table 8.16. This matrix is inverted and the coefficients obtained by the usual process; these results are given in Table 8.17.

The total uncorrected sum of squares is 276, the correction factor for the mean is 240; hence, the total sum of squares is 36 (see Table 8.10). The sum of squares due to the entire model (including μ) is

$$\begin{aligned} SSR &= 0(60) + 0.6(-5.0) + 1.0(5.0)(0.8)(334) \\ &= 269.2 \end{aligned}$$

Hence, by subtraction,

$$SSE = 276 - 269.2 = 6.8$$

which is the same as residual (2) in Table 8.10. The sum of squares for the effects (adjusted) is obtained in the usual fashion:

TABLE 8.16 Example: Reduced Normal Equations

	μ	τ_1	τ_2	x	y
μ	15	0	0	75	60
τ_1	0	10	5	-20	-5
τ_2	0	5	10	-10	5
x	75	-20	-10	445	334

Table 8.17 Example: Inverse and Coefficients

	μ	τ_1	τ_2	x	Coeff.
μ	0.900000	-0.333333	0	-0.166667	0
τ_1	-0.333333	0.266667	-0.066667	0.066667	0.6
τ_2	0	-0.066667	0.133333	0	1
x	-0.166667	0.066667	0	0.0333333	0.8

$$\text{SS(groups)} = [0.6 \quad 1.0]\begin{bmatrix} 0.266667 & -0.066667 \\ -0.066667 & 0.133333 \end{bmatrix}^{-1}\begin{bmatrix} 0.6 \\ 1.0 \end{bmatrix}$$

$$= 12.685$$

and

$$\text{SS}(\beta) = \frac{(0.8)^2}{0.033333}$$

$$= 19.2$$

These results are summarized in the usual analysis of variance format in Table 8.18. It is seen that the test for groups (adj.) is the same test as that for the "standard" covariance analysis, and the test for the effect of the covariate is also the same in the standard covariance analysis (Table 8.14).

The estimated parameters as given in Table 8.16 are *automatically* the adjusted treatment effects; as in Chapter 7,

$$\hat{\tau}_3 = -(\hat{\tau}_1 + \hat{\tau}_2) = -1.6$$

TABLE 8.18 Example: Final Analysis of Variance

Source	d.f.	SS	MS	F
Groups (adj.)	2	12.686	6.343	10.264
Covariance (adj.)	1	19.20	19.200	31.068
Residual	11	6.80	0.618	

The estimated mean $\hat{\mu}$ requires some additional interpretation. Remember that in ordinary multiple regression this term is the y *intercept*, that is, the estimate of μ_x when all values of x are zero; in the linear model (Chapter 7), it is indeed the estimate of the population mean. In this model, the term is more like the y-intercept of regression; hence, estimates of treatment means are only meaningful for a specific value of x; $\bar{x}$ is the value normally used. Thus, the estimated treatment mean (adjusted for covariate) is usually estimated:

$$\hat{\mu}_i = \hat{\mu} + \hat{\tau}_i + \hat{\beta}(\bar{x})$$

Note that the expression

$$\hat{\mu} + \hat{\tau}_i$$

provides estimates of the intercepts in model 2 of Section 8.2.

Confidence intervals on the coefficients as well as treatment effects and adjusted treatment means may be made as in the general linear model, using the expression for the variance of linear functions and the elements of the inverse (Section 7.4).

8.5.2 More General Models

The consideration of the general linear model opens up a wide range of models. Thus, for example, it is possible to consider a model with *two* covariates, say x_1 and x_2, where the coefficient for x_1 is *common* to groups while the coefficient for x_2 is allowed to vary from group to group. For this situation, the model is

$$y_{ik} = \mu + \tau_i + \beta_1 x_{1ik} + \beta_{2i} x_{2ik} + \varepsilon_{ik}$$

which may be rewritten, using dummy variables, as

$$y_{ik} = \mu + \sum \tau_i v_i + \beta_1 x_{1ik} + \beta_{2i} x_{2ik} v_i + \varepsilon_{ij}$$

where

$$v_i = \begin{cases} 1 & \text{if observation in } i^{th} \text{ group} \\ 0 & \text{otherwise} \end{cases}$$

Note that the coefficients β_{2i} actually define an interaction effect. The X'X matrix has the appearance of the matrix in Table 8.14 with the addition of several SS and SP submatrices (one for each group) and corresponding partial sums submatrices added on the right and lower portions. Reductions to adjust for singularities are accomplished by initially redefining the dummy variables as in Section 7.3. Tests of significance for individual coefficients are performed in the usual manner; tests for hypotheses of the type

$$H_0: \beta_{2i} = \beta_{2j} \qquad \text{for all } i \neq j$$

may be performed by a step-down analysis similar to that in Section 8.3 of this chapter (see also Section 8.6).

8.6 A NOTE ON COMPUTATIONS

Although the analyses presented in Sections 8.2 and 8.5 are not conceptually difficult, it is obvious that the associated computations can become exceedingly tedious and unsuitable for hand calculation. Thus, the computations for the models discussed in this chapter are normally performed on high-speed computers. However, direct application of the procedure described in Section 8.2 is not available in most statistical computer packages. The same results can, however, be accomplished with judicious use of available routines augmented by a small amount of hand calculation.

The results of Section 8.2 can be obtained as follows:

1. Perform a single regression on the entire set of data to obtain the statistics for model 1
2. Perform a covariance analysis, which is available in most computer packages, to obtain the results for model 2
3. Perform separate regressions for each of the groups; compute the sum of the residual sums of squares to obtain the error sum of squares for model 3

The results from these three analyses are combined into a single table for the step-down analysis to determine the most appropriate model.

In some computer routines, it is possible to create the appropriate dummy and interaction variables to provide the statistics for the most general linear model described in Section 8.5. Such procedures provide the information for tests of differences in specific coefficients among specific combinations of groups, but do not usually provide the overall tests of differences in the entire regression among all groups. Some computer routines (e.g., procedure GLM of SAS-76) provide for the computation of interactions among continuous and dummy variables and the associated test statistics.

Procedures for performing the regression among means are also not directly available in computer packages. However, this model can be estimated by creating a data set consisting of the means and by performing a weighted regression, which is available in most regression packages. The test of the similarity of the between- and within-regression must, however, be computed by hand or by a program specifically written for that purpose.

8.7 PROBLEMS

1. It is desired to be able to estimate the age of a tree (given by number of rings) from the circumference of the tree. Data on

rings and circumference were recorded on 24 and 25 randomly chosen trees from the south and north slopes, respectively, of some mountains in Colorado. The data are given in Table 8.19. Perform the appropriate analysis to answer the following questions:

TABLE 8.19 Data and Residuals/Yield

	South slope trees		North slope trees	
Obs.	Rings	Circum	Rings	Circum
1	93	33.00	35	20.00
2	164	51.50	30	25.00
3	138	43.10	42	35.00
4	125	23.25	30	17.50
5	129	24.50	21	18.00
6	65	18.75	79	30.25
7	193	43.50	60	28.50
8	68	12.00	63	19.50
9	139	31.75	53	28.00
10	81	20.40	131	52.00
11	73	16.00	155	61.50
12	130	25.50	34	27.25
13	147	44.00	58	23.75
14	51	9.20	55	13.00
15	56	15.40	105	39.25
16	61	6.75	66	24.40
17	115	11.40	70	29.00
18	70	24.75	56	26.25
19	44	8.25	38	10.90
20	44	9.80	43	22.50
21	63	14.30	47	33.25
22	133	31.50	157	65.25
23	239	41.50	100	51.5
24	133	24.50	22	15.6
25	--	--	105	52.0

(a) how well can age be estimated by circumference?

(b) can one equation be used to predict age both on north and south slopes?

2. An important reason for conducting fertilizer response experiments is to make recommendations for optimum fertilizer usage. Data from a 3-year experiment on wheat are given in Table 8.20. Perform a group regression using the model

$$y = \beta_0 + \beta_1 N + \beta_2 N^2 + \beta_3 P + \beta_4 P^2 + \beta_5 NP + \varepsilon$$

Note that levels of N and P are equally spaced, so that orthogonal polynomials may be used.

TABLE 8.20 Data and Residuals Yield

			Year	
N	P	1	2	3
40	0	27	32	42
80	0	33	41	45
120	0	33	45	46
160	0	29	42	49
200	0	30	41	50
240	0	33	44	46
40	40	29	31	41
80	40	29	40	45
120	40	34	39	48
160	40	33	41	45
200	40	33	40	44
240	40	31	38	45
40	80	30	28	40
80	80	32	34	44
120	80	32	41	46
160	80	32	39	43
200	80	32	41	45
240	80	31	35	45

(a) Choose a best model. Interpret findings, using plots if necessary.

(b) Assume that the results of this experiment are to be used to recommend fertilization practices for farmers. Discuss possible limitations and/or qualifications arising from the use of the model(s) you have chosen.

3. It is desired to test the effectiveness of a sales promotion on frozen peas. Eight stores were randomly divided into two groups of four; in one group a 4-week promotion campaign (posters, displays, free recipes, etc.) was conducted; in the other group nothing was done. No price specials (etc.) were offered during the 4-week period. Data were taken on the number of 10-oz. packages (equivalent) sold per week in each of the eight stores (y) and also on the number of customers per store for the week (x).

   (a) Ignore the fact that there are four stores and 4 weeks of treatment and consider the data as two groups of 16 observations each; perform a group-regression analysis.

   (b) Compute a new variable, sales per customer, and perform an analysis to compare the promotion and control groups.

4. Below are the results of a hypothetical experiment on the rate of flow of a liquid through a nozzle (data are coded to make computations easier).

Size of nozzle		
2	4	6
6	4	13
3	6	9
2	8	12
5	11	14
11	8	17
2	14	13

   (a) Do the appropriate ANOVA and take out linear and quadratic contrasts. Interpret.

TABLE 8.21 Data for Problem 3

	Promotion stores							
	1		2		3		4	
Week	x	y	x	y	x	y	x	y
1	409	463	796	958	2067	2051	601	786
2	557	809	880	1219	1502	1947	597	837
3	605	531	901	890	1984	1863	805	733
4	415	563	870	1287	1770	2597	673	1965

	Control stores							
	5		6		7		8	
Week	x	y	x	y	x	y	x	y
1	1305	1000	775	635	143	112	958	826
2	1597	1295	807	608	257	223	1276	1314
3	1201	1193	957	858	307	288	757	531
4	1059	1145	1021	1293	146	152	1159	1400

(b) It is discovered later that the first three observations for each size were obtained using a different type of nozzle than was used for the last three observations. Reanalyze, if necessary, obtaining all relevant contrasts. Interpret results and indicate any differences from part (a).

(c) Can the answer to part (b) be obtained in another way?

APPENDIX

TABLE A.1 Durbin-Watson Statistics:[a] Significance Points of d_L and d_U: 2% (a two-tail test)

	k' = 1		k' = 2		k' = 3		k' = 4		k' = 5	
n	d_L	d_U	d_L	d_U	d_L	d_U	d_L	d_U	d_L	d_U
15	0.81	1.07	0.70	1.25	0.59	1.46	0.49	1.70	0.39	1.96
16	0.84	1.09	0.74	1.25	0.63	1.44	0.53	1.66	0.44	1.90
17	0.87	1.10	0.77	1.25	0.67	1.43	0.57	1.63	0.48	1.85
18	0.90	1.12	0.80	1.26	0.71	1.42	0.61	1.60	0.52	1.80
19	0.93	1.13	0.83	1.26	0.74	1.41	0.65	1.58	0.56	1.77
20	0.95	1.15	0.86	1.27	0.77	1.41	0.68	1.57	0.60	1.74
21	0.97	1.16	0.89	1.27	0.80	1.41	0.72	1.55	0.63	1.71
22	1.00	1.17	0.91	1.28	0.83	1.40	0.75	1.54	0.66	1.69
23	1.02	1.19	0.94	1.29	0.86	1.40	0.77	1.53	0.70	1.67
24	1.04	1.20	0.96	1.30	0.88	1.41	0.80	1.53	0.72	1.66
25	1.05	1.21	0.98	1.30	0.90	1.41	0.83	1.52	0.75	1.65
26	1.07	1.22	1.00	1.31	0.93	1.41	0.85	1.52	0.78	1.64
27	1.09	1.23	1.02	1.32	0.95	1.41	0.88	1.51	0.81	1.63
28	1.10	1.24	1.04	1.32	0.97	1.41	0.90	1.51	0.83	1.62
29	1.12	1.25	1.05	1.33	0.99	1.42	0.92	1.51	0.85	1.61
30	1.13	1.26	1.07	1.34	1.01	1.42	0.94	1.51	0.88	1.61
31	1.15	1.27	1.08	1.34	1.02	1.42	0.96	1.51	0.90	1.60
32	1.16	1.28	1.10	1.35	1.04	1.43	0.98	1.51	0.92	1.60
33	1.17	1.29	1.11	1.36	1.05	1.43	1.00	1.51	0.94	1.59

TABLE A.1 (Continued)

n	k' = 1		k' = 2		k' = 3		k' = 4		k' = 5	
	d_L	d_U	d_L	d_U	d_L	d_U	d_L	d_U	d_L	d_U
34	1.18	1.30	1.13	1.36	1.07	1.43	1.01	1.51	0.95	1.59
35	1.19	1.31	1.14	1.37	1.08	1.44	1.03	1.51	0.97	1.59
36	1.21	1.32	1.15	1.38	1.10	1.44	1.04	1.51	0.99	1.59
37	1.22	1.32	1.16	1.38	1.11	1.45	1.06	1.51	1.00	1.59
38	1.23	1.33	1.18	1.39	1.12	1.45	1.07	1.52	1.02	1.58
39	1.24	1.34	1.19	1.39	1.14	1.45	1.09	1.52	1.03	1.58
40	1.25	1.34	1.20	1.40	1.15	1.46	1.10	1.52	1.05	1.58
45	1.29	1.38	1.24	1.42	1.20	1.48	1.16	1.53	1.11	1.58
50	1.32	1.40	1.28	1.45	1.24	1.49	1.20	1.54	1.16	1.59
55	1.36	1.43	1.32	1.47	1.28	1.51	1.25	1.55	1.21	1.59
60	1.38	1.45	1.35	1.48	1.32	1.52	1.28	1.56	1.25	1.60
65	1.41	1.47	1.38	1.50	1.35	1.53	1.31	1.57	1.28	1.61
70	1.43	1.49	1.40	1.52	1.37	1.55	1.34	1.58	1.31	1.61
75	1.45	1.50	1.42	1.53	1.39	1.56	1.37	1.59	1.34	1.62
80	1.47	1.52	1.44	1.54	1.42	1.57	1.39	1.60	1.36	1.62
85	1.48	1.53	1.46	1.55	1.43	1.58	1.41	1.60	1.39	1.63
90	1.50	1.54	1.47	1.56	1.45	1.59	1.43	1.61	1.41	1.64
95	1.51	1.55	1.49	1.57	1.47	1.60	1.45	1.62	1.42	1.64
100	1.52	1.56	1.50	1.58	1.48	1.60	1.46	1.63	1.44	1.65

TABLE A.1 Durbin-Watson Statistics:[a] Significance Points of d_L and d_U: 10% (a two-tail test)

	k' = 1		k' = 2		k' = 3		k' = 4		k' = 5	
n	d_L	d_U	d_L	d_U	d_L	d_U	d_L	d_U	d_L	d_U
15	1.08	1.36	0.95	1.54	0.82	1.75	0.69	1.97	0.56	2.21
16	1.10	1.37	0.98	1.54	0.86	1.73	0.74	1.93	0.62	2.15
17	1.13	1.38	1.02	1.54	0.90	1.71	0.78	1.90	0.67	2.10
18	1.16	1.39	1.05	1.53	0193	1.69	0.82	1.87	0.71	2.06
19	1.18	1.40	1.08	1.53	0.97	1.68	0.86	1.85	0.75	2.02
20	1.20	1.41	1.10	1.54	1.00	1.68	0.90	1.83	0.79	1.99
21	1.22	1.42	1.13	1.54	1.03	1.67	0.93	1.81	0.83	1.96
22	1.24	1.43	1.15	1.54	1.05	1.66	0.96	1.80	0.86	1.94
23	1.26	1.44	1.17	1.54	1.08	1.66	0.99	1.79	0.90	1.92
24	1.27	1.45	1.19	1.55	1.10	1.66	1.01	1.78	0.93	1.90
25	1.29	1.45	1.21	1.55	1.12	1.66	1.04	1.77	0.95	1.89
26	1.30	1.46	1.22	1.55	1.14	1.65	1.06	1.76	0.98	1.88
27	1.32	1.47	1.24	1.56	1.16	1.65	1.08	1.76	1.01	1.86
28	1.33	1.48	1.26	1.56	1.18	1.65	1.10	1.75	1.03	1.85
29	1.34	1.48	1.27	1.56	1.20	1.65	1.12	1.74	1.05	1.84
30	1.35	1.49	1.28	1.57	1.21	1.65	1.14	1.74	1.07	1.83
31	1.37	1.50	1.30	1.57	1.23	1.65	1.16	1.74	1.09	1.83
32	1.37	1.50	1.31	1.57	1.24	1.65	1.18	1.73	1.11	1.82
33	1.38	1.51	1.32	1.58	1.26	1.65	1.19	1.73	1.13	1.81
34	1.39	1.51	1.33	1.58	1.27	1.65	1.21	1.73	1.15	1.81
35	1.40	1.52	1.34	1.58	1.28	1.65	1.22	1.73	1.16	1.80
36	1.41	1.52	1.35	1.59	1.29	1.65	1.24	1.73	1.18	1.80
37	1.42	1.53	1.36	1.59	1.31	1.66	1.25	1.72	1.19	1.80
38	1.43	1.54	1.37	1.59	1.32	1.66	1.26	1.72	1.21	1.79
39	1.43	1.54	1.38	1.60	1.33	1.66	1.27	1.72	1.22	1.79
40	1.44	1.54	1.39	1.60	1.34	1.66	1.29	1.72	1.23	1.79
45	1.48	1.57	1.43	1.62	1.38	1.67	1.34	1.72	1.29	1.78
50	1.50	1.59	1.46	1.63	1.42	1.67	1.38	1.72	1.34	1.77

TABLE A.1 (Continued)

n	k' = 1		k' = 2		k' = 3		k' = 4		k' = 5	
	d_L	d_U	d_L	d_U	d_L	d_U	d_L	d_U	d_L	d_U
55	1.53	1.60	1.49	1.64	1.45	1.68	1.41	1.72	1.38	1.77
60	1.55	1.62	1.51	1.65	1.48	1.69	1.44	1.73	1.41	1.77
65	1.57	1.63	1.54	1.66	1.50	1.70	1.47	1.73	1.44	1.77
70	1.58	1.64	1.55	1.67	1.52	1.70	1.49	1.74	1.46	1.77
75	1.60	1.65	1.57	1.68	1.54	1.71	1.51	1.74	1.49	1.77
80	1.61	1.66	1.59	1.69	1.56	1.72	1.53	1.74	1.51	1.77
85	1.62	1.67	1.60	1.70	1.57	1.72	1.55	1.75	1.52	1.77
90	1.63	1.68	1.61	1.70	1.59	1.73	1.57	1.75	1.54	1.78
95	1.64	1.69	1.62	1.71	1.60	1.73	1.58	1.75	1.56	1.78
100	1.65	1.69	1.63	1.72	1.61	1.74	1.59	1.76	1.57	1.78

[a]Reprinted by permission of the Biometrika Trustees J. Durbin and G. S. Watson (1951).

TABLE A.2 Orthogonal Polynomials: z-Values[a]

	n = 3		n = 4			n = 5			
	z_1	z_2	z_1	z_2	z_3	z_1	z_2	z_3	z_4
						-2	+2	-1	+1
			-3	+1	-1	-1	-1	+2	-4
	-1	+1	-1	-1	+3	0	-2	0	+6
	0	-2	+1	-1	-3	+1	-1	-2	-4
	+1	+1	+3	+1	+1	+2	+2	+1	+1
$\sum z^2$	2	6	20	4	20	10	14	10	70
λ	1	3	2	1	10/3	1	1	5/6	35/12

	n = 6					n = 7				
	z_1	z_2	z_3	z_4	z_5	z_1	z_2	z_3	z_4	z_5
						-3	+5	-1	+3	-1
	-5	+5	-5	+1	-1	-2	0	+1	-7	+4
	-3	-1	+7	-3	+5	-1	-3	+1	+1	-5
	-1	-4	+4	+2	-10	0	-4	0	+6	0
	+1	-4	-4	+2	+10	+1	-3	-1	+1	+5
	+3	-1	-7	-3	-5	+2	0	-1	-7	-4
	+5	+5	+5	+1	+1	+3	+5	+1	+3	+1
$\sum z^2$	70	84	180	28	252	28	84	6	154	84
λ	2	3/2	5/3	7/12	21/10	1	1	1/6	7/12	7/20

TABLE A.2 (Continued)

	n = 8					n = 9				
	z_1	z_2	z_3	z_4	z_5	z_1	z_2	z_3	z_4	z_5
	-7	+7	-7	+7	-7					
	-5	+1	+5	-13	+23					
	-3	-3	+7	-3	-17					
	-1	-5	+3	+9	-15	0	-20	0	+18	0
	+1	-5	-3	+9	+15	+1	-17	-9	+9	+9
	+3	-3	-7	-3	+17	+2	-8	-13	-11	+4
	+5	+1	-5	-13	-23	+3	+7	-7	-21	-11
	+7	+7	+7	+7	+7	+4	+28	+14	+14	+4
$\sum z^2$	168	168	264	616	2184	60	2772	990	2002	468
λ	2	1	2/3	7/12	7/10	1	3	5/6	7/12	3/20

	n = 10					n = 11				
	z_1	z_2	z_3	z_4	z_5	z_1	z_2	z_3	z_4	z_5
						0	-10	0	+6	0
	+1	-4	-12	+18	+6	+1	-9	-14	+4	+4
	+3	-3	-31	+3	+11	+2	-6	-23	-1	+4
	+5	-1	-35	-17	+1	+3	-1	-22	-6	-1
	+7	+2	-14	-22	-14	+4	+6	-6	-6	-6
	+9	+6	+42	+18	+6	+5	+15	+30	+6	+3
$\sum z^2$	330	132	8580	2860	780	110	858	4290	286	156
λ	2	1/2	5/3	5/12	1/10	1	1	5/6	1/12	1/40

TABLE A.2 (Continued)

	n = 12					n = 13				
	z_1	z_2	z_3	z_4	z_5	z_1	z_2	z_3	z_4	z_5
						0	-14	0	+84	0
	+1	-35	-7	+28	+20	+1	-13	-4	+64	+20
	+3	-29	-19	+12	+44	+2	-10	-7	+11	+26
	+5	-17	-25	-13	+29	+3	-5	-8	-54	+11
	+7	+1	-21	-33	-21	+4	+2	-6	-96	-18
	+9	+25	-3	-27	-57	+5	+11	0	-66	-33
	+11	+55	+33	+33	+33	+6	+22	+11	+99	+22
$\sum z^2$	572	12012	5148	8008	15912	182	2002	572	68068	6188
λ	2	3	2/3	7/24	3/20	1	1	1/6	7/12	7/120

	n = 14				
	z_1	z_2	z_3	z_4	z_5
	+1	-8	-24	+108	+60
	+3	-7	-67	+63	+145
	+5	-5	-95	-13	+139
	+7	-2	-98	-92	+28
	+9	+2	-66	-132	-132
	+11	+7	+11	-77	-187
	+13	+13	+143	+143	+143
$\sum z^2$	910	728	97240	136136	234144
λ	2	1/2	5/3	7/12	7/30

TABLE A.2 (Continued)

	n = 15				
	z_1	z_2	z_3	z_4	z_5
	0	-56	0	+756	0
	+1	-53	-27	+621	+675
	+2	-44	-49	+251	+1000
	+3	-29	-61	-249	+751
	+4	-8	-58	-704	-44
	+5	+19	-35	-869	-979
	+6	+52	+13	-429	-1144
	+7	+91	+91	+1001	+1001
$\sum z^2$	280	37128	39780	6466460	10581480
λ	1	3	5/6	35/12	21/20

	n = 16				
	z_1	z_2	z_3	z_4	z_5
	+1	-21	-63	+189	+45
	+3	-19	-179	+129	+115
	+5	-15	-265	+23	+131
	+7	-9	-301	-101	+77
	+9	-1	-267	-201	-33
	+11	+9	-143	-221	-143
	+13	+21	+91	-91	-143
	+15	+35	+455	+273	+143
$\sum z^2$	1360	5712	1007760	470288	201552
λ	2	1	10/3	7/12	1/10

[a]Reprinted from Anderson and Houseman (1942) by permission of the Iowa Agriculture and Home Economics Experiment Station.

BIBLIOGRAPHY

Alldredge, J. R., and N. S. Gilb (1976), Ridge Regression; An Annotated Bibliography, *Int. Stat. Rev. 44*:335-360.

Anderson, R. L., and E. E. Houseman (1942), *Tables of Orthogonal Polynomial Values*, Iowa State Experiment Station, Bulletin 297.

Anscombe, F. J. (1960), Rejection of Outliers, *Technometrics, 2*: 123-147.

Bhat, U. N. (1972), *Elements of Applied Stochastic Processes,* Wiley, New York.

Box, G., and G. Jenkins (1976), *Time Series Analysis, Forecasting and Control* (Rev. ed.), Holden-Day, San Francisco.

Brownlee, K. A. (1960), *Statistical Theory and Methodology in Science and Engineering*, Wiley, New York.

Draper, N. R., and H. Smith (1966), *Applied Regression Analysis*, Wiley, New York.

Durbin, J., and G. S. Watson (1951), Testing for Serial Correlation in Least Squares Regression II, *Biometrika 25:*173-175.

Graybill, F. (1976), *Theory and Application of the Linear Model*, Duxbury Press, North Scituate, Massachusetts.

Hamilton, T. H., and I. Rubinoff (1963), Species Abundance: Natural Regulation of Insular Variation, *Science 142*(3599):1575-1577.

Hohn, F. E. (1964), *Elementary Matrix Algebra* (2nd ed.), Macmillan New York.

Horst, P. (1963), *Matrix Algebra for Social Scientists*, Holt, Rinehart and Winston, New York.

Johnston, J. (1972), *Econometric Methods* (2nd ed.), McGraw-Hill, New York.

LaMotte, L. R., and R. R. Hocking (1970), Computational Efficiency in the Selection of Regression Variables, *Technometrics 12*: 83-94.

Malinvaud, E. (1966), *Statistical Methods of Econometrics*, Rand McNally, Chicago.

Mallows, C. L. (1973), Some Comments on Cp, *Technometrics 15*:661-666.

Morrison, D. F. (1976), *Multivariate Statistical Methods* (2nd ed.), McGraw-Hill, New York.

Ostle B., and R. W. Mensing (1975), *Statistics in Research* (3rd ed.), Iowa State University Press, Ames.

Rao, P. (1971), Some Notes on Misspecification in Multiple Regression, *Statistician 25*:37-39.

Searle, S. (1971), *Linear Models*, Wiley, New York.

Snedecor, G. W., and W. G. Cochran (1967), *Statistical Methods* (6th ed.), Iowa State University Press, Ames.

Snee, R. D. (1973), Some Aspects of Nonorthogonal Data Analysis, *Journal of Quality Technology 5*:67-69.

Southern Cooperative Series, Bulletin 101, October 1965.

U. S. Department of Commerce (1976), Business Conditions Digest.

U. S. Department of Commerce (1966-1968), Weekly Weather Bulletin, Volumes 53-55.

Valavanis, S. (1959), *Econometrics*, McGraw-Hill, New York.

Walls, R. C., and D. L. Weeks (1969), A Note on the Variance of a Predicted Response in Regression, *Statistician 23*(3):24-26.

INDEX